LOCOMOTION PAPERS   LP246

The
The Ramsey North Branch

by
Peter Paye

THE OAKWOOD PRESS

© Oakwood Press & Peter Paye 2020

Published by Oakwood Press, an imprint of Stenlake Publishing Ltd, 2020

British Library Cataloguing in Publication Data
A Record for this book is available from the British Library
ISBN 978 0 85361 744 0

Printed by Claro Print, Office 26/27, 1 Spiersbridge Way, Glasgow, G46 8NG.

All rights reserved. No part of this book may be reproduced or transmitted in any form or by any means, electronic or mechanical, including photocopying, recording or by any information storage and retrieval system, without permission from the Publisher in writing.

By the same author and published by Oakwood Press:

The Snape Branch (2005)
The Hadleigh Branch (2006)
The Jersey Eastern Railway (2007)
The Framlingham Branch (2008)
The Wisbech & Upwell Tramway (2009)
The Bishop's Stortford, Dunmow and Braintree Branch (2010)
The Mellis & Eye Railway (2012)
The Aldeburgh Branch (2012)
The Hayling Railway (2013)
The Ely & St Ives Railway (2014)
The Axminster & Lyme Regis Light Railway (2015)
The Saffron Walden Branch (2017)
The Ramsey East Branch (2018)

Title page: LM class '4MT' class 2-6-0 No. 43146, passing Peterborough North signal box with the late-running Ramsey North branch freight train on a dull November day in 1961.
Author

Rear cover: An extract from the 1930 one inch Ordnance Survey map showing the route of the Ramsey North branch (not reproduced to scale). *Crown Copyright*

Oakwood Press, 54-58 Mill Square, Catrine, KA5 6RD,
Tel: 01290 551122 *Website:* www.stenlake.co.uk

Contents

	Introduction	5
Chapter One	The Struggle for a Railway	6
Chapter Two	Great Eastern Takeover	19
Chapter Three	Pre-Grouping Days	33
Chapter Four	Grouping to Closure	55
Chapter Five	The Route Described	75
Chapter Six	Permanent Way, Signalling and Staff	89
Chapter Seven	Timetables and Traffic	100
Chapter Eight	Locomotives and Rolling Stock	121
Appendix One	Lengths of Platforms, Sidings, etc.	171
Appendix Two	Bridges and Culverts	172
Appendix Three	Level Crossings	173
	Acknowledgements	174
	Bibliography	175
	Index	176

Railway Enthusiasts' Club 'Charnwood Forester' railtour train hauled by 'C12' class 4-4-2T No. 67380, pauses at St Mary's for a photographic stop on 14th April, 1957. Note the ornate cast post supporting the level crossing gate on the up side of the line.

J. Spencer Gilks

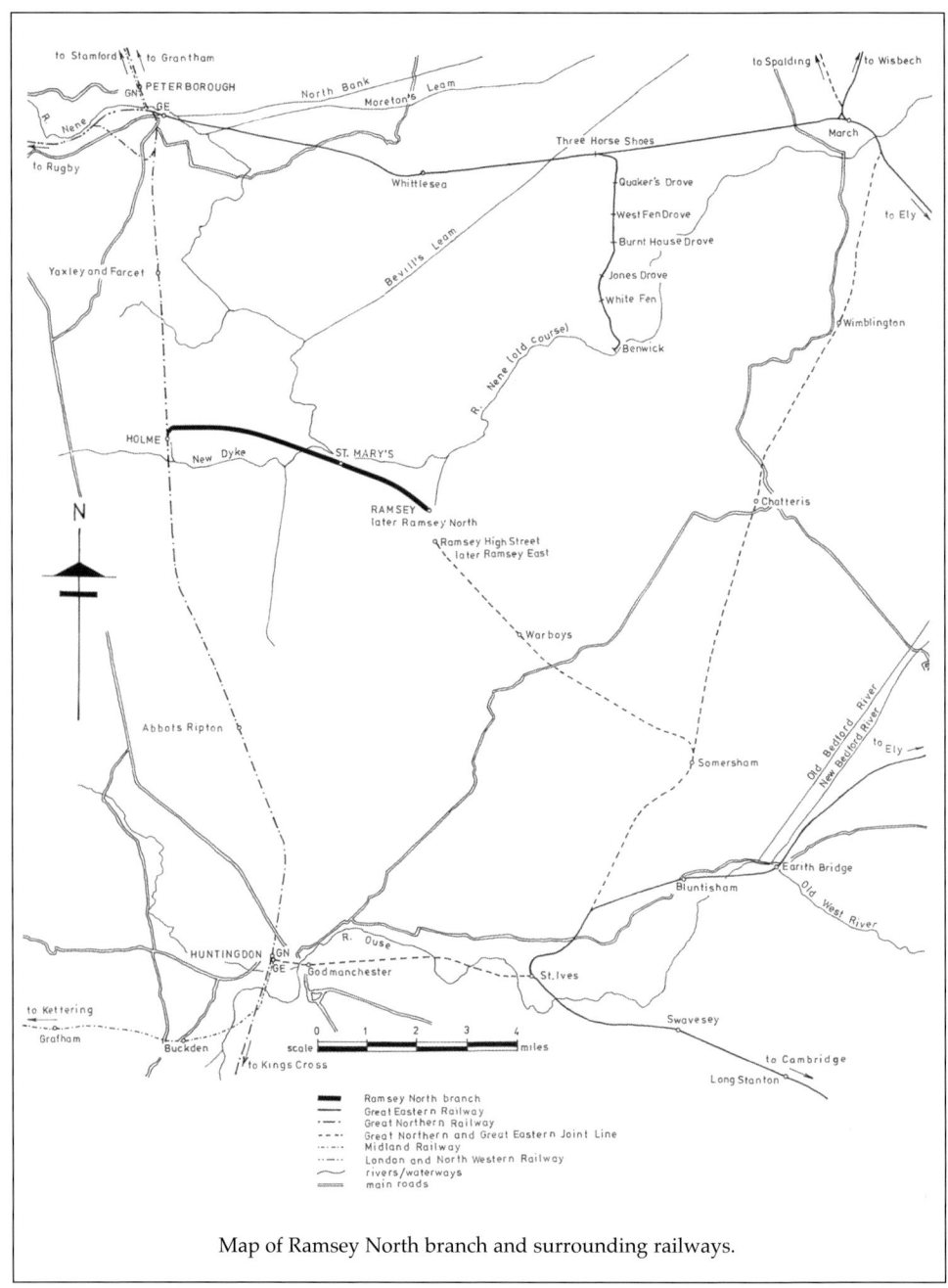

Map of Ramsey North branch and surrounding railways.

Introduction

The small fenland market town of Ramsey in Huntingdonshire, was one of the many which suffered from the agricultural depression in the mid-19th century. With the coming of the railways the town was stranded between the Great Northern Railway (GNR) to the west and the Eastern Counties Railway (ECR) to the east. After various unsuccessful attempts had been made to place Ramsey on the railway map, local businessmen and landowners promoted their own line to link up with the GNR at Holme, six miles to the west. The Ramsey Railway, authorized in 1861 and opened two years later, proved to be an anomaly. Initially worked by the GNR, the majority of shares of the local company were quickly acquired by the Great Eastern Railway (GER), successors to the ECR, to prevent their arch rivals from using the line as a stepping stone into East Anglia. From 1875 the Ramsey Railway was absorbed by the GER, which then immediately leased the branch to the GNR. This arrangement remained in force until the 1923 Grouping and, therefore, this erstwhile GER line never saw a GER locomotive or train on its metals.

Unsuccessful attempts to extend the branch beyond Ramsey to St Ives and the failure to unite with the later Ramsey & Somersham Junction Railway, opened in 1889, sealed the Ramsey North line to a domiciled fate. Receipts from the sparsely-populated area were poor and, except for Huntingdon and Peterborough market days, the trains carried few passengers. By the late 1920s bus services had infiltrated into this area of the Fens, offering almost door-to-door transport for local people. It was therefore of no surprise when the London & North Eastern Railway (LNER) authorities withdrew all passenger services after the 10.15 am train as an economy measure during the depressed days of 1931. The few remaining passenger services finally succumbed in 1947 but freight, always the more lucrative for traffic receipts, remained intact for a further two decades until it too dwindled with the transfer of many items to road hauliers. The freight trains were finally withdrawn from the branch in 1973.

An attempt has been made at tracing the history of the line. Details have been checked with documents that are available, but apologies are offered for any errors which might have occurred.

Peter Paye,
Bishop's Stortford

Note: The railway spelling of Whittlesea (with an 'a') instead of the geographical spelling of Whittlesey, has been used throughout the text for clarity.

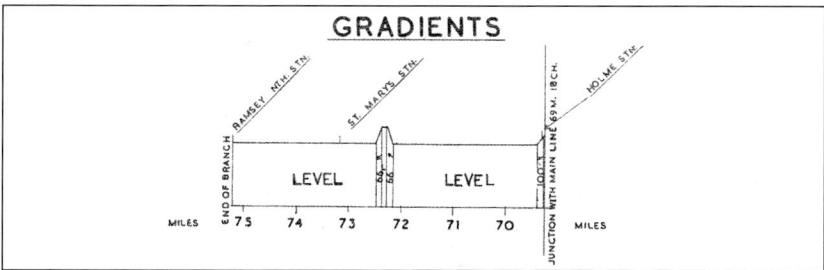

Chapter One

The Struggle for a Railway

For centuries the dank Fenland covering an area of 680,000 acres, encompassing sections of Lincolnshire, Cambridgeshire, Huntingdonshire and Norfolk, was isolated from the surrounding counties and few strangers dared venture into its interior. Whilst the silt fens of the north were uniformly flat, the peat fens of Cambridgeshire and Huntingdonshire contained many fen islands, some rising to over 100 ft above sea level. Fenland settlements, such as Ely and St Ives were established on these pockets of high land and from these ridges Hereward the Wake fought the Norman invaders. From Roman times waterways and causeways were utilized as routes between settlements, and the basis of such a transport system rapidly led to a spread of population. Near the western edge of the Fens at Ramsey, the foundations of an abbey were laid in AD969. The name was derived from two English words Ram and Eie or Eye, which compounded means Island of the Ram. The Benedictine Abbey dedicated to Our Lady, St Benedict and All Holy Virgins, and initially staffed by a prior and 12 monks, was built above the flood plain. By the 12th century the establishment was one of the largest and richest in the country and had gathered a small community adjacent to the Abbey. A market town developed during the following century and in 1247, Henry III granted authority for the holding of an annual fair. Most of the magnificent Aabbey buildings were destroyed soon after the reformation and in 1665 the plague devastated the town, when over 400 people died. It was thought the infection had been conveyed from London in cloth tailored for a coat. In the 18th century the town again suffered two disastrous fires, which destroyed most of the wooden and thatch buildings.

Despite such setbacks, by the early 19th century Ramsey was one of the four largest communities in the former administrative county of Huntingdon. With the drainage of the Fens, initially engineered by the Dutchman, Sir Cornelius Vermuyden, in the early 17th century, the surrounding land was gradually drained utilizing both steam and wind power, allowing smallholdings and farms to be established on the newly acquired soil. Ramsey relied almost entirely on waterways for connections with the surrounding area and at one time the thoroughfare, known as Great Whyte, was a navigable channel through the town centre. Land drainage brought about the re-routing of such waterways and although barge and lighter traffic provided cheap conveyance for goods, it was slow and ponderous. The primitive road systems connecting Ramsey to St Ives, Huntingdon and Peterborough were also established but often during wet or wintry weather routes were impassable because of flooding.

In the 1840s and 1850s the area suffered the agricultural depression. Ramsey was facing economic stagnation unless valuable crops grown in the locality could be quickly dispatched to markets to obtain the best possible prices. With roads and waterways unable to provide such a service the remedy lay with the hope of attracting the railway to the town, but this was easier said than done.

Several unsuccessful schemes to construct a railway from London to York were mooted before the advent of the Great Northern Railway came with the publication of ambitious plans in 1844. After many trials and tribulations the promoters successfully obtained the necessary Parliamentary approval on 26th June, 1846 to build the line. The works were to be financed by the raising of share capital of £5,600,000 and borrowing powers of £1,868,000. The initial contracts for the construction of the Peterborough to Maiden Lane, London section were awarded in November 1848 but progress was slow despite the employment of between 5,000 and 6,000 men. Funds were difficult to obtain and major works, including nine tunnels, Welwyn viaduct and the three-mile crossing of the treacherous Whittlesea Mere, effectively delayed opening of the railway until 7th August, 1850, followed by the extension into King's Cross on 14th October, 1852. The railway by-passed the town of Ramsey, six miles to the west at Holme, where the GNR provided a station as a railhead for the district, much to the disappointment of the inhabitants.

Unfortunately Ramsey had also suffered a similar fate with railway routes approaching from the east. By 16th May, 1842 the Northern & Eastern Railway (N&E) had extended the 28 miles from Stratford to Bishop's Stortford and in the following year sought powers to extend a further 10 miles to Newport. In an agreement dated 23rd December, 1843 the N&E was taken over by the Eastern Counties Railway on a 999 years lease, with effect from 1st January, 1844. The line was subsequently extended from Bishop's Stortford through Newport and Cambridge to Brandon via Ely to join up with the Norfolk Railway route from Norwich and opened on 30th July, 1845. In the same year the ECR obtained powers to build a branch from Chesterton, two miles north of Cambridge, to St Ives and this line opened to traffic on 17th August, 1847, the same day as the Ely & Huntingdon company opened their railway from St Ives to Godmanchester. Meanwhile in 1846, the Wisbech, St Ives & Cambridge Junction Railway obtained powers to connect the towns mentioned and to carry cattle and corn from St Ives market to the port of Wisbech. The railway was opened north of March on 3rd May, 1847 and between March and St Ives on 1st February, 1848, under the control of the ECR, which had absorbed the smaller company. Stations were provided at Somersham and Chatteris to serve the district, but both were over seven miles from Ramsey.

As early as 1845 the first of many schemes was made to route a railway through Ramsey. In that year plans were submitted for a line connecting Huntingdon and Wisbech from a south-facing junction off the proposed London & York Railway, in the parish of Great Stukeley and passing via Ramsey, Warboys and Chatteris, then on to March and Wisbech. Alarmed by such proposals and fearing intrusion into their territory, the ECR countered the threat by publishing plans to construct a branch to Ramsey from Somersham on the proposed Wisbech, St Ives & Cambridge Junction Railway. The necessary plans for the Eastern Counties Extension Railway, running for a distance of 6 miles, 5 furlongs and 9 chains, were submitted to the Clerk of the Peace for Huntingdonshire on 30th November, 1846 and the Private Bill Office of the House of Commons on 23rd December. Both parties encountered vehement opposition from local landowners and found difficulty raising the capital, the original scheme was allowed to lapse, whilst the latter was rejected in the 1847 session of Parliament.

No further positive efforts were made to place Ramsey on the railway map for 13 years, until proposals were published on 9th November, 1859 for the Somersham, Ramsey & Holme Railway. Commencing 10 chains north of the ECR station at Somersham, the proposed line passed through the parishes of Pidley-cum-Fenton, Warboys, Wistow, Bury, Ramsey, Upwood, Wood Walton and Connington to terminate at a junction with the GNR at Holme station. The cost of the line was estimated at £60,000, with an additional £15,000 for purchase of land and legal expenses. The embryonic company was empowered to enter into a working agreement with the ECR or GNR or both. The plans were deposited at the Parliamentary Private Bill Office on 23rd December, 1859 and it was confidently forecast that the proposed line would carry thriving passenger traffic, serving as it did an area with a combined population of 14,000 people. Substantial goods traffic was also envisaged with the transit of corn, potatoes, coal, iron, timber, bricks and artificial manures. The promoters optimistically forecast the railway would provide better communication than the existing fen waterways, where routes were circuitous and slow. Initially a petition was lodged against the Bill by the Middle Level Drainage Commissioners, pending confirmation of sureties for the protection of their navigation, but this was withdrawn on 27th March, 1860 and at the beginning of April the Somersham to Holme Railway Bill was placed in the list of unopposed proposals. The situation was encouraging but difficulties were being experienced raising capital. The proposed contractors, Simpson & Walker of Park Farm, Ely, had offered to take capital of £25,000, some of it in shares whilst others had promised to subscribe a total of £20,000. The undertaking could take advantage of borrowing powers of £26,600 but this at best still left a deficit of £3,400. Approaches were made to both the ECR and GNR for financial support but both declined to assist in any way, and the promoters were faced with the inevitable decision of abandoning the scheme. A meeting was subsequently held at Ramsey on 5th May, 1860 chaired by the local MP and extensive landowner, Edward Fellowes. Despite the pressure by some saying the scheme was still viable, Fellowes counteracted the optimistic belief and by majority vote the Somersham to Holme undertaking was abandoned. .

The collapse of this scheme failed to deter local businessmen in Ramsey who were anxious for a railway to serve the town and surrounding district. Energetic canvassing produced enough support for a line to Holme, and on 7th November, 1860 plans were published, and duly submitted to the Clerk of the Peace for Huntingdonshire on 30th November, and the parish authorities and Parliamentary Bill Office on 23rd December, 1860. The need for a railway to increase trade and eliminate economic stagnation was evident for a correspondent in the *Huntingdon, Isle of Ely, Bedford, Peterborough and Lynn Gazette* reported:

> Ramsey is in a dreadful state, the poor are nearly all out of work and have no food, hundreds of them are standing in the streets day after day, while there is no subscription afoot for their relief and there is not a single shilling being given away. What is to be

done? Are our fellow creatures to die? Perhaps if you ask the public it might do good - for things cannot be much worse.

Unfortunately the situation did deteriorate, adverse weather in the spring of 1861 caused a dearth of crops and arable farming suffered. Some root vegetables were ploughed in and fields were freshly sown with spring corn. Many acres remained devoid of crops.

The Bill for the proposed railway enjoyed an uneventful passage through Parliament, and with the Great Northern Railway showing a close interest in the new line, the local promoters asked if the main line company would subscribe to the capital and work the railway. The GNR Directors duly discussed the matter at their Board meeting on 25th May, 1861 and advised the Ramsey Railway of their inability to subscribe. The General Manager was, however, requested to negotiate a working agreement.

The Ramsey Railway Act 1861 (24 & 25 Vict. Cap. cxciv) which received the Royal Assent on 22nd July, 1861, authorized the company to build a railway, commencing from a junction with the GNR at or near Holme station, in the parish of Holme-cum-Glatton, otherwise Holme Fern in the county of Huntingdon and passing through the parishes of Holme and Ramsey to terminate in a field owned by Thomas Darlow, near Ramsey Gas Works. Two years were allowed for the compulsory purchase of land and three years for the completion of the works. To provide the necessary finance for building the railway, the statute authorized the raising of £30,000 in £10 shares with additional borrowing powers of £10,000, once half of the original capital was paid up. The company was authorized to provide level crossings across road No. 11 in the parish of Holme and road No. 83 in the parish of Ramsey. By clause 21 the company was not permitted to shunt trains over level crossings or allow trains to block such level crossings. The following clause required the provision of lodges for crossing keepers or stations to be erected adjacent to each public level crossing. Because of the extensive network of drainage channels crossed by the proposed railway, the company was empowered to seek the authority of the Bedford Level Corporation and the Middle Level Drainage Commissioners before commencing construction, and complete such works to their satisfaction. The railway company was restricted from flushing away or eroding earth or soil within 20 yards of the landside of any bank, and any land taken by the company was subject to drainage taxes. Clause 34 of the Act specifically stated the Ramsey Railway was to be carried over the River Nene and Hooks Lode or New Dyke by good substantial bridges with the soffit on the underside of the bridge not less than 16 feet clear height from the Middle Level Datum Line, each one span being not less than 30 ft wide.

The initial Board of Directors appointed by the Ramsey company, were Arthur Ballantine, Chairman, Thomas Darlow and John Poulter. Other interested parties included Ibbertson Saunders and Isaac Palmer. Frederick W. Thorp of St Ives was Solicitor to the company. Darlow was the chief instigator in providing Ramsey with a gas supply and latterly a new water pump at a cost of £15 for the local fire brigade. Edward Fellowes, after his reluctance to back the previous scheme, initially played no part in the new undertaking,

ANNO VICESIMO QUARTO & VICESIMO QUINTO

VICTORIÆ REGINÆ.

Cap. cxciv.

An Act to authorize the Construction of a Railway from *Holme* to *Ramsey* in the County of *Huntingdon*. [22d *July* 1861.]

WHEREAS the Construction of a Railway from the *Holme* Station of the *Great Northern* Railway to *Ramsey* in the County of *Huntingdon* would be of local and public Advantage: And whereas Plans and Sections of the Railway, showing the Line and Levels thereof, with a Book of Reference to the Plans, containing the Names of the Owners, Lessees, and Occupiers of the Lands through which the said Railway will pass, have been deposited with the Clerk of the Peace for the County of *Huntingdon*: And whereas the Persons herein named, with others, are willing at their own Expense to carry the said Undertaking into execution, but they cannot do so without the Authority of Parliament: May it therefore please Your Majesty that it may be enacted; and be it enacted by the Queen's most Excellent Majesty, by and with the Advice and Consent of the Lords Spiritual and Temporal, and Commons, in this present Parliament assembled, and by the Authority of the same, as follows:

1. "The Companies Clauses Consolidation Act, 1845," "The Lands Clauses Consolidation Act, 1845," "The Lands Clauses Consolidation Acts 8 & 9 Vict. cc. 16., 18., & 20., and

[*Local*.]   32 D

First page of the Ramsey Railway Act of 22nd July, 1861.

prompting adverse comments in the local press but he had added his full support in the weeks before the passing of the Act.

At an open meeting held at Ramsey on 27th July, 1861, the assembly was informed that work on the railway would commence as soon as possible after the first Directors' meeting and once a suitable contractor had been engaged. Satisfied with their progress, the Directors then decided on the possible extension of the railway from Ramsey via Warboys and the Hursts to St Ives, and Sir Charles Fox, the company Engineer was requested to commence the necessary surveys with a view to presenting a Bill to Parliament for the 1862/63 session. Fox had trained as a surgeon before joining the Liverpool and Manchester Railway, where he learned the technicalities of railway construction. He then set up in the contracting business and was knighted for his work in erecting the Crystal Palace in 1851 and for its subsequent removal to Sydenham. He was later involved with other schemes, including the Hayling Railway.

The first ordinary meeting of the shareholders was held at the Mechanics Institute, Ramsey, at 3 pm on Wednesday 25th September, 1861. Frederick Thorp, the company Solicitor advised the gathering of the encouraging support, both financial and moral, for the new line and gave details of the terms of the working agreement with the GNR. The Directors had full confidence in the soundness of the undertaking and expressed, 'it will prove a great boon to the district, and of a remunerative character to the shareholders'. Edward Fellowes was then elected to the Board of Directors, whilst William Orris and Edward Beecheno were appointed company auditors, at a salary of five guineas per annum.

During the late summer and autumn of 1861 little was achieved. Sir Charles Fox quickly made the preliminary surveys for the extension and reported no untoward opposition preventing construction. The initial Board, was, however, tiring of their responsibilities and no effort was made to generate enthusiasm. By 2nd December, 1861, the Directors decided work on the construction on the authorized line would be postponed until the New Year because of the impending winter weather and their inability to find a suitable contractor. At the same meeting Thomas James, a solicitor from Ramsey was appointed as the company Secretary, at a sum not exceeding £100 for the first year.

Early in the New Year the ineffective Board received a shake up when Edward Fellows MP, took over as Chairman and proceeded to show his enthusiasm and authority. At the second meeting of ordinary shareholders held at the Institute Room at Ramsey on 12th March, 1862, Fellowes announced that £7,090 capital had been raised, offset by expenditure of £1,114 4s. 10d. on Parliamentary and other expenses. Sir Charles Fox advised that after surveying to St Ives, he had set out the course of the authorized line, prepared the necessary plans, sections and estimates for the contractor and negotiated favourable terms for the purchase of land with most of the landowners. Suitable sites were established for stations at St Mary's and Ramsey, and negotiations were underway with the GNR General Manager regarding the junction, station accommodation and interchange facilities at Holme. Fellowes apologized to the shareholders for the slow progress, blamed partly on their inexperience in

THE GREAT NORTHERN RAILWAY COMPANY

AND

THE RAMSEY RAILWAY COMPANY.

Agreement for working the Traffic of the Ramsey Railway by the Great Northern Railway Company.

AN AGREEMENT made the 1st day of July, 1862, between the GREAT NORTHERN RAILWAY COMPANY (hereinafter called the Great Northern Company) of the one part, and the RAMSEY RAILWAY COMPANY (hereinafter called the Ramsey Company) of the other part. WHEREAS the Ramsey Company was incorporated by the Ramsey Railway Act 1861, for the purpose of making a railway to commence by a junction with the main line of the Great Northern Railway, at or near the Holme station of that railway, in the parish of Holme, in the county of Huntingdon, and to terminate in the parish of Ramsey, in a field near the gas works, also in the said county of Huntingdon. And whereas the said Ramsey Company are about to complete the said railway, together with the necessary stations at Ramsey and St. Mary's respectively. And whereas the Great Northern Company have agreed with the Ramsey Company to work the traffic of the said Ramsey Railway when and so soon as the same shall be completed and fit to be opened for public traffic, and for the term of seven years next thereafter; and for the purpose of carrying out the agreement the said two Companies have agreed to enter into this contract. Now these presents witness that each of the said two Companies, parties hereto, do hereby for themselves, their successors and assigns, covenant and declare with, and to the other of them, their successors and assigns, in manner and to the effect following:— *Ramsey Act of Incorporation.*

1. That the said Ramsey Company shall complete their railway, and the junction with the Great Northern Railway at Holme, together with the necessary junction signals, station signals, stations, gate lodges, residences for gatekeepers at such level crossings as may be necessary, and all such sidings as are now agreed to by the Great Northern Company *Ramsey Company to complete line.*

First page of the working agreement between
the Great Northern Railway and the Ramsey Railway Company.

building a railway but more importantly because of the tardy response of shareholders providing the finance for construction. He concluded on an optimistic note, however, saying everything was ready for a contractor to commence work and it was hoped to open the line within a year.

In April 1862 the services of a contractor were finalized and William Smith Simpson, of Park Farm, Ely was awarded the contract at a cost of £26,512, inclusive of all works and stations. Simpson and his partner Walker had been the proposed contractors for the failed 1859 Somersham, Ramsey and Holme scheme, and were also involved with the construction of the Ware, Hadham & Buntingford Railway in Hertfordshire and the West Norfolk Railway. Despite poor work on both lines, later in 1864 they were also awarded the contract for the construction of the Ely, Haddenham & Sutton Railway. Wet weather at the end of April and in early May 1862 precluded the commencement of construction works and the continuous rain concluded with extensive flooding of land near Ramsey. At a poorly attended meeting of shareholders on 12th May, many voiced their anxiety at the continuing delay and lack of finance, as an additional £10,000 was required to complete the line. Fellowes countered the pessimism by announcing he was personally subscribing half of the amount, whilst another £2,600 had been offered by various individuals. At a subsequent Board meeting, held at the offices of Sir Charles Fox, Spring Gardens, London on 20th May, the Directors reiterated their determination to proceed with construction without delay. Thorp, the Solicitor was instructed to complete the contract with W.S. Simpson so that the first sod could be cut within a fortnight. The ceremony was duly conducted at the beginning of June with the Honorable Mrs Fellowes, wife of the Chairman, performing the actual deed.

After exhaustive negotiations, the agreement with the GNR to work the traffic on the Ramsey Railway for an initial period of seven years was finalized on 1st July, 1862, and ratified on 22nd of the month. The terms included:

1. The Ramsey Railway to complete the railway and junction with the Great Northern Railway at Holme, together with any necessary junction signals, also station signals, stations, gates and gate lodges and residences for gate keepers, and all such sidings as agreed by the GNR. A line of telegraph was also to be erected and, if the Board of Trade inspector stipulated, engine turntables provided at each end of the railway.
2. The station at Ramsey to be constructed with a residence for a clerk-in-charge as well as complete station accommodation. The station at St Mary's to be constructed with residences for a clerk-in-charge and a gatekeeper.
3. The junction with the GNR at Holme to be constructed at the expense of the Ramsey Railway.
4. The Ramsey Railway to provide additional stations and/or sidings, if required.
5. The GNR to have sole working of traffic, and if the line was later constructed with double track or extended to St Ives, no other company was to have access to or use Holme station.
6. The maintenance of the permanent way to be the responsibility of the Ramsey Railway for the first six months and thereafter maintenance to be carried out by the GNR.

Construction of the railway finally commenced in June 1862 but was initially delayed by a dearth of local labour, which was involved with the harvest. The attractions of wages slightly above agricultural rates lured many away from the fields and Simpson's men then made steady progress across the difficult and peaty subsoil, where a base of brushwood, sand and hardcore was used. In order to expedite completion and make up for lost time the 200 or so contractor's men, consisting of local men augmented by Scottish and Irish navvies, worked seven days a week. By September 1862, 1,269 shares had been purchased representing a shareholding of £12,690. The continuation of construction on Sundays shocked the religious zealots of the neighbourhood, for in November 1862 the local paper reported the 'sorrow of seeing the desecration of the Sabbath by men working on that day. We would rather defer the opening of the railway than this should be!'

In the meantime, parties surveying a proposed line from Ramsey to Chatteris for the recently formed Great Eastern Railway had visited the area. The GER had been created by the amalgamation of the ECR and other East Anglian railways from 1st August, 1862 and indeed the GER General Manager had reported on 29th October to his Traffic Committee that it was essential for the protection of GER interests to extend their line to Ramsey 'which could be at a very moderate cost, not exceeding £40,000, as the district is level'. The proposed railway left the St Ives to March line at or near the 83 mile post and passed through the parishes of Chatteris, Doddington, Warboys, Wistow, Bury and Ramsey to terminate at an end-on junction with the authorized Ramsey Railway in or near a field in the occupancy of Thomas Darlow, at a point 50 yards from the gas works at Ramsey. Local opinion was, however, supportive of the St Ives line, as it was considered the ECR had neglected the area by withdrawing support from the earlier 1859 scheme for the railway linking Somersham with Ramsey and Holme. Eight days after the GER announced their Chatteris and Ramsey proposals, the plans for the Ramsey to St Ives line were published on 18th November. Copies of both schemes were submitted to the Parliamentary Private Bill Office on 23rd December, 1862, the Ramsey Railway Directors seeking authority to build a railway commencing from an end-on junction with the authorized Ramsey Railway in field No. 54 in the parish of Ramsey and passing through the parishes of Bury, Wistow, Warboys, Old Hurst, Wood Hurst and St Ives, to terminate at a junction with the GER St Ives to Wisbech line at or near the coal depot occupied by Thomas Coote, near St Ives station.

Within a month the GER authorities were showing interest in the St Ives proposal, as it would save unnecessary expenditure if that scheme were adopted instead of the Chatteris project, and negotiations commenced with the Ramsey Railway Board. Having entered into a working agreement with the local company the GNR authorities were highly suspicious of the move, especially when the Ramsey Directors advised the GNR Assistant General Manager that the GER was to abandon their Ramsey to Chatteris scheme and allegedly subscribe half of the capital cost of the St Ives line. The GNR Directors, well aware of old hostilities, requested an update of the state of relationships between the rival factions. Another concern of the GNR Traffic Manager was the poor patronage of Holme station. The village, with a population of 644 souls, hardly justified the provision of a station, although originally it was hoped to attract traffic from a wide catchment area including the town of

Ramsey. Unfortunately the shortest route between Ramsey and Holme station was by water and very few people adopted this method of transport before embarking on a train journey. The situation was partially resolved when the GNR Engineer advised his Board on 15th January, 1863 that the local authority proposed to build a road from Yaxley to Holme. As it was optimistically envisaged people would travel to Holme to join the train instead of journeying to Peterborough the GNR was asked if they would support the scheme by offering financial backing. The Directors, keen to see an increase in traffic at the station, advised the Engineer on 4th February that, subject to satisfactory agreement between both parties, the railway company was willing to assist.

Inclement weather in the first weeks of the new year hampered construction of the Ramsey Railway, but work proceeded as and when possible. On Friday 30th January, 1863 a local navvy, William Bridgefoot, Junior, was killed by a ballast wagon, which passed over him when he fell across the rails. At the inquest held at Ramsey on Saturday 31st January, it was established that the deceased was employed with other navvies unloading sand from wagons on to the permanent way formation. Bridgefoot was in charge of one of the horses pulling the wagons but progress was slow and in an endeavour to pick up speed the horse was whipped. After repeated flogging the animal became restless and gathered pace. Bridgefoot lost control of the horse, stumbled and fell across the rail between horse and wagon. Unfortunately the wet sand prevented the navvy from dodging aside and the wagon passed over him with fatal results. Recording a verdict of accidental death, the coroner commented on the fateful events affecting the Bridgefoot family. On the day prior to the accident Bridgefoot's brother died of a fever, whilst a sister was stricken with the same illness. Another brother had also recently broken a leg.

Improvements in the weather enabled Simpson's men to accelerate construction work and by mid-March most of the formation was laid out and permanent way in position to within two miles of the junction at Holme. Work on the stations was carried out concurrently with the track work and the timber station at Hern (St Mary's) was almost completed by local contractor Isaac Bateman. Unfortunately the smooth progress of work was marred by an accident to Henry Dawson, a local 21-year-old who was employed as a navvy on the construction work. During earth-moving operations where track was laid, he slipped and fell across the rail where a wagon passed over both feet. Despite severe bleeding, he was taken to Peterborough hospital for treatment and subsequently had one foot amputated.

Meanwhile the Ramsey and St Ives promotion was finding difficulty engendering sufficient local enthusiasm and a claim was made to the GER authorities for financial backing. After lengthy discussion, during which the abandonment of the Chatteris scheme was agreed, the GER Directors advised their Secretary on 5th March, 1863 to reply to the St Ives promoters saying that if their Act was passed, the GER was prepared to advance a sum of not exceeding £30,000. Flushed with success, Fellowes as Chairman of the Ramsey & St Ives scheme, then wrote to the GNR proposing that that company subscribe £36,000 or half of the share capital to the undertaking. If this proved acceptable the promoters promised to enter into a working agreement with the

GNR when the railway was completed. The GNR Board at King's Cross, already assured of traffic from Ramsey at almost minimal cost, was not desirous of incurring further financial involvement and bluntly turned down the suggestion on 30th April, 1863.

During the spring of 1863 work on the Ramsey Railway continued and by mid-April track was laid to the terminus. The arrival of the embryonic railway proved a great attraction to the inhabitants of Ramsey who flocked to the site of the station to watch with great interest the activities of the Engineer and navvies. The interest was further increased when the townsfolk were offered unofficial rides in the horse-drawn open wagons. However, on the evening of Friday 24th April tragedy struck. Trucks of ballast being drawn by horses were being shunted in the future station area when a Mrs Richardson and her children asked for a ride in an open wagon, from which the ballast had recently been discharged. After a short ride the wagon was halted some 50 yards from the end of the line, where the mother and children were asked to alight. Unfortunately a little girl, 7-year-old Rebecca, thought she could ride for a further distance and clambered back on to the vehicle just as the short train of wagons started to move. Missing her footing she slipped and before anyone noticed her movements, two trucks passed over her with tragic consequences. At the subsequent inquest held before Thomas James, the coroner and also Secretary of the railway company, no blame was attributed to the driver who was unable to see the child as he reined the horse to start the wagons. A verdict of accidental death was duly recorded.

At the 'very thinly attended' fourth ordinary general meeting of the Ramsey Railway, held in the Institute Rooms at Ramsey on 30th April, 1863, the Engineer advised that the railway would be completed and ready for opening in six weeks. The track, ballasted with ashes rather than granite to give a lighter base, was laid from Ramsey to within half a mile of the junction at Holme. All culverts and bridges were completed and level crossings and fences in position. The engine shed, goods shed, weighbridge and coal wharf at Ramsey had been erected whilst St Mary's station was virtually complete. Despite all calls being made on shareholders the company was still in debt to a total of £1,209 and, because of the advanced state of construction and the amount owing to Simpson, the gathering agreed to the company taking advantage of the £10,000 borrowing powers authorized by the empowering Act. At the conclusion of the meeting, the Directors, Engineer and shareholders travelled by horse-drawn wagons as far as the line was completed, a mile short of Holme station, to view the construction work.

Unable to raise the required capital, support for the Ramsey & St Ives scheme waned, much to the disappointment of the Ramsey Board and shareholders, and the undertaking was withdrawn at the Parliamentary committee stage. The GER Chatteris Bill, which had also been advanced, was rejected by Parliament in the same session. Because of the rapid advancement of the Holme to Ramsey line, Sir Charles Fox persuaded the Ramsey Railway Board to persevere with an extension, but this time from Ramsey to Somersham. How the Directors and shareholders hoped to succeed when the line to St Ives had failed is puzzling, but between his commitment to the Ramsey Railway and other schemes, Fox duly conducted another survey of the proposed route.

By mid-May construction of the authorized line was almost completed except for minor adjustments to track ballasting. On 23rd May, 1863 notice was given to the Board of Trade for the necessary inspection and the local newspaper commented, 'as soon as approval is given the line will be opened'. Having taken the decision to seek borrowing powers for £10,000 as authorized by the Act, Edward Fellowes, as Chairman of the Ramsey line wrote to the General Manager of the GNR requesting that company to guarantee interest on the debenture issues. The GNR Board tersely declined assistance advising, 'The company had no legal powers to guarantee bonds of another company'.

Colonel W. Yolland carried out the official Board of Trade (BoT) inspection on 8th June, 1863. Travelling by special train across the branch and stopping at various places en route, the inspector found that land had been purchased for double track with the formation on average 17 ft wide but only a single track railway, 5 miles 58½ chains in length, was offered for inspection. Bridges had been constructed for a single track railway only and sidings were laid at Holme, St Mary's and Ramsey. After close examination of the permanent way, Yolland turned his attention to the two underbridges on the line. The first structure, with three openings, giving a total span of 34 ft 7 in. was constructed of cast-iron girders resting on timber piles, whilst the other of two small openings was built entirely of timber. After testing, both were considered sufficiently strong, giving only slight deflections under load. The inspector found the fencing of the Ramsey line 'somewhat peculiar', consisting mainly of ditches with a quick growing fence to be planted at the proper season of the year. Both the Ramsey Railway and GNR officials advised Yolland that this was sufficient for the district and a considerable proportion of the GNR main line in the area was similarly fenced. The inspector was not impressed and required the company to supply additional post and rail fencing along the line, to stop cattle straying on to the railway. On finding no turntable at Ramsey, Yolland was advised the GNR proposed to work the line as a branch with 'One Train Only' working utilizing a tank locomotive, and therefore no turning was required. Learning of the possibility of extending the railway beyond Ramsey, the Inspector insisted that if this was achieved, then a turntable was to be located at Holme.

At the conclusion of his visit, Yolland found the following required attention. Fang bolts were to be added on the inside of the rails on the sleepers next to the joints within a period of three months. Clocks were to be provided at stations and placed where they could be seen from the platform. The locomotive coke stage at Ramsey was placed too near to the track and required moving back, whilst the facing points at all stations required to be locked. The Colonel also stipulated that care be taken when an engine or train worked round the 13 chain radius curve at Holme. Because of the peaty nature of the subsoil on which the railway was built, additional ballasting was required to prevent sagging of the formation. The Inspector refused to sanction the opening of the line to passenger traffic until the fang bolts were in position and 'an undertaking had been given by the Ramsey Railway of the early completion of the other outstanding works'.

The disappointed Directors immediately requested Sir Charles Fox as their Engineer and Simpson, the contractor, to bring the line up to the required standard. Navvies were soon reballasting and remedial work, except for the fencing was quickly completed. The Ramsey Railway Directors, however, objected to paying for the erection of the fencing and Fellowes advised the GNR General Manager of the situation and asked if the railway could still be opened. On 30th June, the General Manager sought guidance from his Directors asking if the line could be opened on and from 6th July, 1863. The GNR Directors, whilst requesting further information, tentatively agreed to the line opening, provided the Ramsey company absolved the GNR for loss or claim arising from any accident caused by the lack of fencing.

In the meantime the local press were critical of the lack of opening:

> Rumours of the proposed opening of the railway have been extensively current during the last few days, and much disappointment has been felt by the Ramsey people on account of their not being permitted to avail themselves of the completed line to have attended the Peterborough Choral Festival and the Hunts Flower Show held last week. We trust the inauguration of so important an epoch in our local history will be celebrated in a matter befitting the occasion.

Embarrassed by local public criticism Simpson arranged to provide at his own expense, and thereby not contravening the BoT embargo on public opening, a special private train to Peterborough Fair on 10th July. All persons travelled free of charge and the children were reported to be 'highly delighted with their visit'.

Compromise finally came on 14th July after correspondence between the GNR General Manager and Sir Charles Fox. It was agreed after legal consultation and correspondence with the BoT that the line could open for traffic provided the local company fenced the line. Messrs Faber and Astell of the GNR were asked to negotiate with the Ramsey Railway Directors and the latter, faced with no alternative, agreed to fence the railway. Simpson, having obtained the material, set his men to construct the fencing at strategic positions leaving the remaining sections to a later date. This appeared to satisfy the BoT and the GNR Directors agreed to open the line to traffic.

After advertisements appeared in local papers on 18th July, 1863, the GNR district officers working to a pre-arranged plan sent a locomotive and rolling stock to Ramsey on 21st July to inaugurate the service, and the line was opened to the public with the departure of the 8.05 am train to Holme on Wednesday 22nd July, 1863. Only a few passengers bothered to travel on the first train and the *Huntingdon, Bedford, Cambridge and Peterborough Gazette* reported the opening as 'a tame affair', with 'no public dinner and no single hurrah'. Another newspaper reported about 100 people gathered at Ramsey station to watch the first departure, which carried 30 passengers. The engine was 'tastefully decorated and most profusely adorned with evergreens interspersed with flowers' whilst the carriages were adorned with a few sprigs of evergreen. The GNR initially provided a service of six passenger trains each way, weekdays only, but this proved over generous and by 1st August was reduced to five each way, with the withdrawal of the 2.25 pm Ramsey to Holme train and the return working.

Chapter Two

Great Eastern Takeover

The opening of the line coincided with the first day of the Ramsey annual cattle fair, which lasted overall for three days. As well as trading in cattle, the local populace combined the event with an excuse for merrymaking and entertainment. The railway company fully expected rich pickings as animals were conveyed to and from the fair but the Directors were doomed to disappointment, for the *Peterborough Advertiser* reported fewer stalls than in previous years and of those that attended most were of 'decidedly inferior character'. To make matters worse none of the usual sideshows were in attendance.

Two months after the opening of the railway, Edward Fellowes reported to the half-yearly meeting of Ramsey Railway shareholders, held on 22nd September, 1863, that traffic receipts were satisfactory and calls on shares were far better 'now that shareholders could see the fruits of their investments in operation'. The Board hoped the line would develop a 'fair amount of traffic' and, whilst not paying a high rate of dividend, would still yield a fair return. The outstanding debts amounting to £437 would be paid off within a few weeks. Local farmers had already complained of a lack of a goods shed at St Mary's, while others felt the conveyance of corn traffic would be greatly enhanced if tramway links were made between the various stations and local waterways, as grain conveyed by fen lighter and barge could be transferred into railway wagons for faster transit to markets.

When the line opened to traffic, a somewhat piecemeal approach to the collection and delivery of parcels had been adopted by the railway company, but as traffic increased it became evident that definitive arrangements would have to be negotiated. After several enquiries, the GNR authorities announced on 3rd November, 1863, that Mr Butler of Ramsey had been appointed agent for the collection and delivery of parcels in the town.

Earnings for the period from the opening of the line to 31st December, 1863 were encouraging, as the figures below show:

	£.	s.	d.
Passenger	255	5	0
Horses	2	6	2
Parcels	7	19	0
Merchandise	326	19	10
Livestock	6	0	9
Minerals	72	16	6
Sundries	1	10	0
Total	672	17	3

Early in the new year the contractor responsible for conveying Her Majesty's Royal Mail from Huntingdon to Ramsey gave notice of withdrawal as he was losing money on the operation. Evidently by this time the railway was providing a faster parcels service from Ramsey to Huntingdon and London, for the Post Office authorities found it difficult to find a replacement contractor.

A further attempt for an approach to Ramsey from the east, came with the publication of plans for a railway from St Ives, 9 miles 4 furlongs and 3½ chains in length. Although plans were deposited with the Clerk of the Peace for Huntingdonshire on 28th November, 1863 nothing came of the venture. The GER authorities were also showing increasing concern as to their inability to connect the town of Ramsey with their system and thereby thwart a possible extension of the GNR into its East Anglian territory by using the Ramsey Railway. In the meantime Sir Charles Fox had completed the survey for the extension of the Ramsey Railway to Somersham. To the GER this was tantamount to open warfare and the Board quickly initiated a blocking action. In March 1864, the company announced its own plans to build a line from Somersham to Ramsey. The Ramsey Railway Board was alarmed at the effect such a railway would have on their proposed extension line. After an emergency Board meeting Mr Jordan, Parliamentary agent for the local company asked the GNR authorities assistance in opposing the GER scheme, as it would have the effect of rendering their line and station at Ramsey useless. The matter was raised at the GNR Board meeting at King's Cross on 22nd March, 1864, when the Ramsey Railway request for £600 to assist the Parliamentary Committee costs was agreed. The application was, however, later declined leaving the local company no alternative but to accept the development of the GER proposals.

The decision by the GNR Directors to absolve themselves from assisting the Ramsey Railway in defending its territory, combined with a negative attitude adopted in operating the branch, gave cause for concern in the local company boardroom. The GER authorities, however, were waiting developments and when the unwillingness of the GNR became known, approaches were made to Edward Fellowes to reach a compromise. The Ramsey Company was obviously unable to finance the extension of their line to Somersham without a great deal of difficulty. Equally the GER wished to safeguard their territory against possible incursion by a foreign railway company, and although the GNR had shown no intention of late to use the Ramsey line for such purposes, the threat was always there whilst the local railway company remained nominally independent. The GER Directors acted swiftly and suggested a takeover of the Ramsey company.

The absorption of the Ramsey Railway by the GER, however, required an Act of Parliament, to which the GNR would vehemently object. After protracted negotiations it was agreed the GER would offer to purchase for £40,000, all the shares of the Ramsey Railway and transfer them to a GER nominated holding. For any shares not transferred, the GER promised to pay interest at five per cent per annum. The main line company was at that time experiencing financial difficulties and in order to raise the necessary capital, was forced to borrow £40,000 from the Union Bank. The negotiations were made without the knowledge of the majority of Ramsey Railway shareholders and it was therefore of great surprise to the gathering at the half-yearly meeting held at the Institute Rooms, Ramsey on 6th April, 1864, that Fellowes announced the sale of the undertaking to the GER. Fellowes also advised that no dividends were available as outstanding accounts with the GNR and the contractor were not settled. Traffic had not fully developed as was 'to be expected after so short a time of operation' and money was still owed to the contractor.

The transfer of the Ramsey Railway shares to the GER was agreed at the end of April and on payment of £36,000 the majority of the holding was duly registered in the names of GER Directors, James Goodson and Colonel Jarvis, leaving Edward Fellowes, as Chairman of the Ramsey Railway and his partner W. Stafforth holding the remaining 122 shares. Having painlessly acquired the majority of the shares of the minor railway, and safeguarded their own territory without serious objections from the GNR, the GER Directors strove to consolidate their acquisition and decided to seek Parliamentary approval for outright purchase.

At the same time the plans for the railway linking Somersham and Ramsey were prepared for enactment, the route being surveyed by Charles R. Chaffins, with Robert Sinclair of the GER as Engineer. Notice of the Bill advocating the purchase of the Ramsey Railway by the GER and the extension line to Somersham were duly deposited with the Clerk of the Peace for Huntingdonshire on 29th November, 1864 and the House of Commons Private Bill Office on 24th December.

Receipts for the year 1864 showed an initial reduction during the first six months but this was more than offset by an increase in the second half of the year when root vegetables and wheat harvested from the surrounding fenland was dispatched to market.

Six months ending	30th June, 1864			31st December, 1864		
	£	s.	d.	£	s.	d.
Passenger	198	12	8	269	10	8
Horses	4	5	0	1	12	11
Parcels	11	10	8	13	17	6
Merchandise	315	8	8	471	4	5
Livestock	9	13	6	6	18	7
Minerals	45	18	11	76	8	10
Sundries		1	10	0	0	0
Total	585	11	3	839	12	11

January 1865 was unusually cold in the fens and during the month several of the rivers were frozen to a considerable depth. Fen skating was a popular pastime for the active of Ramsey and district when skates were at a premium. Operation of the branch was also affected during this period when points froze solid on at least two occasions, whilst several trains were delayed as the primitive locomotive allocated to work the line slipped to a standstill because of the frost-covered rails.

The attempt by the GER to purchase the Ramsey Railway outright was thwarted early in March 1865, when the Committee of the House of Commons rejected the Bill after receiving objections from the GNR. In recompense, however, the preamble of the Bill for the railway linking Somersham with Ramsey received the necessary consent. In spite of this minor setback, Edward Fellowes was keen to rid himself of the responsibility of the Holme to Ramsey railway and wrote to Bishopsgate on 27th April, 1865 asking if the GER would purchase the outstanding 122 shares, accepting at the same time the expenses incurred by the Ramsey Railway from 1st April, 1864. As compensation Fellowes offered to arrange for the GER to receive the proportion of receipts paid by the GNR to the local company. The GER was now in an invidious position, for having had their case for takeover thrown out at Parliamentary level, they considered it undesirable to absorb all the shares. Fellowes was subsequently advised that the GER would take 22 of the outstanding shares leaving the Ramsey Railway as a nominally independent concern with the Chairman Edward Fellowes and Director W. Stafforth holding 50 shares each. The GER Directors also advised that it was necessary to keep the Ramsey Railway operating and in the event of the GNR refusing to work the services, the GER would seek to operate the line.

The branch again suffered from adverse weather conditions on Monday 22nd May, 1865, when an electrical storm and torrential rain passed across the district around Ramsey and Huntingdon. The lightning was so severe that several buildings were struck. The heavy rain that followed fell so quickly that the ballast was washed from under the track in several places on the branch, and train services were halted until the line was inspected and remedial repairs made.

On 2nd June, 1865 the extension line was duly authorized by the Great Eastern Railway (Ramsey Branch) Act 1865 (28 Vict. cap. lxii). Commencing at a point 25 chains or thereabouts westwards of the mile post marked 77 miles on the St Ives to March line at Somersham and leaving that line by a curve of 2 furlongs 25 chains radius, the proposed railway ran for a distance of 7 miles, 7 furlongs and 8 chains to terminate by a junction with the Ramsey Railway at a point 12 chains or thereabouts north of the School Drove Ground, where the same was crossed on the west by the said railway. Three years were allowed for the compulsory purchase of land and five years for the completion of works. The extension proposal made purely as a safeguard against possible GNR intrusion into GER territory, was allowed to lapse almost as soon as the ink was dry on the statute. The Bishopsgate company showed no inclination to proceed with construction and the reneging aroused strong local feelings of resentment.

The annual Ramsey Fair, which commenced on Saturday 22nd July, 1865, and continued until the following Wednesday, attracted a considerable number of visitors to the town. The GNR took advantage of earning additional revenue and offered cheap day return tickets to Ramsey from Holme, Huntingdon and Peterborough.

Just when receipts for livestock traffic were increasing to a satisfactory level, a severe outbreak of cattle plague hit the counties of Huntingdonshire, Bedfordshire and Cambridgeshire. Many thousands of animals were slaughtered and livestock movements were barred. From the end of October 1865 until early in the New Year no livestock was conveyed on the branch freight services and even when restrictions were lifted the traffic took some months to recover. Despite this setback receipts continued to improve, much to the satisfaction of the Ramsey Railway Board and the GNR authorities. The reliance on agricultural traffic for the major proportion of receipts was again evident as the figures for the second half of the year were greater.

Six months ending	30th June, 1865			31st December, 1865		
	£	s.	d.	£	s.	d.
Passenger	189	6	2	268	0	8
Horses	4	0	10	0	0	0
Parcels	13	11	5	19	15	8
Merchandise	386	6	0	522	8	9
Livestock	13	6	4	17	14	9
Minerals	56	0	8	44	12	2
Sundries	6	4	0	8	1	0
Total	668	15	5	880	13	0

The first serious incident on the branch occurred on the evening of Friday 14th December, 1866. After carrying out repairs to the line, a platelayer was working a trolley along the track when the last down train from Holme to Ramsey approached in the darkness. Hearing the train, the platelayer jumped off the trolley and ran back along the branch, shouting to warn the driver. The noise of the locomotive and the wind sweeping across the fens made his voice inaudible and the engine ploughed in to the trolley, demolishing it completely. After bringing the train to a halt, the driver learned the truth of the matter from the shocked platelayer. No damage was sustained by the locomotive or coaching stock, whilst the passengers only suffered a minor shaking.

Traffic receipts for 1866 continued to show an increase over the previous year although most of the enhancement was attributed to the conveyance of agricultural products. Passenger increase was minimal whilst minerals showed a decline especially when compared with 1864.

Six months ending	30th June, 1866			31st December, 1866		
	£	s.	d.	£	s.	d.
Passenger	198	0	1	278	1	6
Parcels	18	18	3	19	13	11
Merchandise	411	7	1	583	9	10
Livestock	20	18	5	16	12	11
Minerals	58	14	6	47	12	9
Sundries	8	1	0	8	1	0
Total	715	19	4	953	11	11

Another accident, with more serious consequences occurred on 16th February, 1867 at Holme. Head porter Donald McKay was deputed to take a hamper destined for Biggleswade from the Ramsey train, across the platform to the London train. Two minutes later he was discovered seriously injured between the platform edge and the wheels of the brake van. The next passenger train to Peterborough was stopped especially to convey the injured man to hospital but he died on the journey.

On 9th April, 1867, Fellowes again approached the GER Board asking if they would take the remaining Ramsey Railway shares and adopt the line. The Directors declined such action and asked for details of the traffic receipts to 31st March, 1867. Until the advent of the GNR and the building of the branch, the inhabitants of Ramsey knew little of the village of Holme as the main route of access between the two settlements was by water. The regular interchanging of passengers at the junction brought a closer tie between the communities so that when the villagers of Holme arranged the ceremony of the laying of the foundation stone at their new Wesleyan Chapel on 26th September, 1867, the GNR arranged for many people from Ramsey to travel to the ceremony by strengthening the branch train with additional coaches and providing a cheap return fare.

In 1867 the upswing of traffic receipts continued with passenger and parcels showing a minor increase, which was partially offset by a reduction of earnings on merchandise, especially during the second half of the year. The largest increase came with the conveyance of mineral traffic as coal and coke merchants in the area finally placed almost total reliance on the railway for the conveyance of their traffic.

Six months ending	30th June, 1867			31st December, 1867		
	£	s.	d.	£	s.	d.
Passenger	216	17	11	287	9	6
Parcels	19	18	10	21	4	1
Merchandise	463	15	11	559	8	9
Livestock	22	3	9	10	9	1
Minerals	73	1	4	92	13	8
Sundries	2	1	9	1	12	10
Total	797	19	6	972	17	11

Traffic receipts on the Ramsey Railway exceeded all expectations in 1868. Lower receipts on livestock and minerals compared with the previous year were more than offset by vastly improved takings on passenger, parcels and merchandise traffic especially during the second half of the year.

Six months ending	30th June, 1868			31st December, 1868		
	£	s.	d.	£	s.	d.
Passenger	226	2	7	323	5	10
Parcels	23	0	8	27	16	11
Merchandise	435	16	7	757	0	1
Livestock	15	4	1	8	2	9
Minerals	63	5	4	79	3	1
Sundries	1	1	10	1	10	1
Total	764	11	1	1,196	18	9

On 10th April, 1869, Edward Fellowes wrote from 3 Belgrave Square, London to the GER Directors reminding them of the obligation their company had made in taking the shares of the Ramsey Railway. He also complained of the services provided by the GNR. In alluding to the 1862 agreement between the GNR and the Ramsey Railway, the main line company had undertaken to keep the Ramsey and Holme railway and stations in good repair but despite several complaints to King's Cross, the woodwork on the stations and structures was already 'rotting for the want of painting'. As the working agreement with the GNR had almost expired the Ramsey Railway Chairman asked what action he should take to get the GNR to paint the stations. Fellowes again reminded the GER Board of the Ramsey Railway intention of extending their line to Somersham, an application that had later been withdrawn from the Private Bill Office, when the arrangement had been made between the two companies for the GER to pay £40,000 for the total shareholding or five per cent per annum on all shares not transferred. He also expressed profound disappointment to the proposed abandonment of the railway from Ramsey to Somersham authorized by the 1865 Act and enquired of the Directors as to whether the GER ever intended to extend the Ramsey Railway to Somersham. Fellowes, still desperate to rid himself of the responsibility of the branch, also again asked if the outstanding shares held by W. Stafforth and himself could be taken by the GER.

On receipt of the letter at Bishopsgate the GER authorities sought further information and the subject was discussed at the Board meeting on 22nd April, 1869. The Secretary reported that the capital of the Ramsey company totalled £30,000 shares and £10,000 bonds. To finance the transfer the GER had borrowed £40,000 from the Union Bank in April 1864. Of this total, 1,088 Ramsey Railway shares had been transferred to Captain Jarvis for £10,880, whilst 1,790 shares were held in the name of James Goodson for £17,900. Twenty-two £10 shares were later taken but held in no name. Edward Fellowes and W. Stafforth, the Ramsey Railway Directors, held the 100 remaining shares. The £10,000 bonds were held solely in the name of James Goodson. The transaction had cost the GER £3,338 18s. 5d. in interest payments. The gross receipts of the Ramsey line, after removal of the GNR working expenses, totalled £2,259 9s. 7d., and less the £317 13s. 7d. office expenses for the period 1st April, 1864 to 31st March, 1867, left a net amount of £1,941 16s. 0d., of which £618 3s. 1d. had been credited to the GER capital account and the remaining £1,323 12s. 11d. to the revenue account.

After considering the future role of the Ramsey Railway in the GER network, the Secretary advised Fellowes that he should, as a start, enforce the GNR to paint the stations as required by the working agreement. The GER Chairman also reiterated that because of monetary problems, the extension from Ramsey to Somersham could not be considered until a future date and therefore it was not thought necessary to take over the outstanding 100 shares of the local company.

The Ramsey Railway thus continued as a self-contained branch line and as the operation was proving a liability, little was done to attract additional trade by offering attractive rates except when absolutely necessary. The GER authorities considered full absorption of the company could not be entertained until the initial seven year working agreement between the local company and the GNR was due for renewal. In pursuance of this arrangement, Fellowes continued to forward the Ramsey Railway proportion of receipts received from the GNR, less administrative expenses, to the GER at Bishopsgate. Such was the reluctance of the King's Cross management to pay promptly, that only on 28th May, 1869 did Fellowes forward a cheque to the GER for £736 13s. 0d. as the balance for the 31st March, 1868. Fellowes advised the GER Directors that no account for traffic receipts for the year ending 31st March, 1869 had yet been received from the GNR.

In the meantime, because of the impending date for the renewal of the working agreement, the GNR authorities conducted their own investigation into the working of the Ramsey line. This revealed that the number of passengers travelling from the branch stations during the months of September and October 1869 and the proportion of receipts divided between the Great Northern and Ramsey Railway (RR) were:

Destination	Total Passengers	£	s.	d.	Proportion payable to GNR £	RR £
Huntingdon	504	44	5	0	27	17
King's Cross	303	114	8	0	105	9
Peterborough	2,211	142	10	6	78	64
Local	2,719	44	3	2	0	44

Expenditure in the period to 31st October, 1869 revealed why the GNR was showing a reluctance to work the line, for the branch was operating at a loss.

	£	
Maintenance of way and works and stations	455	equal to 6.9d. per train mile
Provision of locomotive power	616	equal to 9.34d. per train mile
Carriage and wagon repairs	132	equal to 2.00d. per train mile
Traffic expenses		
Salaries and wages	322	
Fuel, lighting etc.	39	
Greasing	10	
Clothing	7	
Printing, stationery, tickets etc.	6	
Horse shunting	39	
Junction expenses	70	
	493	equal to 7.48d. per train mile
General charges		
Average on the cost of the GNR expenses exclusive of electric telegraph and Railway Clearing House expenses	34	
Electric Telegraph	32	
Railway Clearing House	23	equal to 1.35d. per train mile
Rates and taxes	66	equal to 1.00d. per train mile
	155	
Total	1,851	equal to 28.07d. per train mile
Proportion allocated to GNR for working expenses	913	equal to 13.08d. per train mile
Loss to GNR	938	

Receipts for the year ending 31st October, 1869 and the apportionment of takings to each company was as follows:

	Gross Receipts	Apportionment to	
		GNR	RR
	£	£	£
Passenger traffic	548	274	274
Parcels traffic	32	16	16
Merchandise	1,094	547	547
Livestock	20	10	10
Coal	132	66	66
Total	1,826	913	913

These receipts were equal to 27.69d. per train mile

The receipts for 1869 showed a decline of £12 from the 1868 totals and again revealed the reliance on agricultural traffic for almost two-thirds of the earnings.

Six months ending	30th June, 1869			31st December, 1869		
	£	s.	d.	£	s.	d.
Passenger	238	6	2	308	0	10
Parcels	22	1	5	25	9	8
Merchandise	472	2	2	727	18	0
Livestock	12	5	4	5	4	2
Minerals	58	13	10	76	14	9
Sundries	8	3	0	1	11	7
Total	811	11	11	1,144	19	0

In the new year Edward Fellowes was still concerned over the indifferent attitude of the GER regarding the future of the Ramsey Railway, and the understandably reluctant attitude of the GNR, now that the working agreement of 1862 was close to expiry. Despite the large shareholding of the GER in the smaller railway, the Ramsey Directors were still unhappy with the abandonment of the Somersham scheme. Fellowes, perplexed by the attitude taken at Bishopsgate wrote to his fellow MP, Sir Edward Watkin asking if the Ramsey and Somersham railway was abandoned, would the GER take over the Holme and Ramsey line and work the railway with their own locomotives or alternatively by hiring engines from the GNR. Sir Edward duly broached the subject at the GER Board meeting on 31st March, 1870 but received little satisfaction. The Directors duly noted the concerns of the Chairman of the Ramsey line and curtly advised that abandonment of the Somersham scheme was proceeding:

> If Mr Fellowes and his followers wished to pursue the making of a railway on to Somersham and provide two thirds of the cost of a cheaply built line, the Directors would recommend the GE shareholders pay the remainder and after work the line at a cost, subject to arbitration.

Sir Edward reported the decision to Fellowes and his co-Director the following day. After further discussion at local level withdrawal of the opposition to the abandonment of the GER Ramsey to Somersham scheme was agreed, together with promises for the future safe working of the Holme to Ramsey line. On 14th April, 1870 Sir Edward Watkin, representing the GER, and Edward Fellowes signed an agreement between the two companies stating that all future opposition by the Ramsey Railway to the GER abandonment scheme would be withdrawn and a new Bill to Parliament be supported, advocating the following:

1. Petitions in support of any future application to the vesting of the Holme and Ramsey Railway in the GER be agreed.
2. The GER undertake to work at least as efficiently as at present the Holme and Ramsey line.
3. If in the next or any following sessions of Parliament until 1872, local interests project a cheap extension railway from the Holme to Ramsey line to connect with the GER system and provide half the capital, the GER Board will recommend their proprietors to provide the remaining capital, not exceeding £25,000, and will work the extension so as not to cause a loss to the GER.

On 19th May, 1870 the GER Directors agreed to place the proposals to the next meeting of the GER proprietors. Despite the offer made in clause 2 of the agreement, the GER authorities had no intention of operating the line and with the working agreement expiry date of 30th June, 1870 looming closer, the main line company authorities remained aloof. However, at the behest of some of the GER Directors, Edward Fellowes finally broached the subject and wrote to King's Cross on 21st May, 1870 asking if the GNR would continue to work the line from 1st July on the existing terms. On 10th June, 1870 the GNR Board appointed Colonel Duncombe and Mr Waterhouse to negotiate on their behalf with full powers to settle.

The Great Eastern Railway (General Powers) Act 1870 (33 and 34 Vict. cap. xxxvi) which received the Royal Assent on 20th June, 1870 finally sanctioned the abandonment of the construction of the railway authorized by the Great Eastern Railway (Ramsey Branch) Act of 1865.

After an initial meeting with Fellowes, the GNR agreed to work the Ramsey line for a fortnight on the same terms, pending an early meeting with the GER as to the future working arrangements. The GER authorities appeared reluctant to negotiate with their King's Cross counterparts and matters were delayed until Friday 8th July when the parties finally consulted. No outright decision was made and the GNR representatives agreed to continue working the line for two weeks subject to an early settlement. The GER contingent returned to Bishopsgate to consult with their senior officers as to whether they should work the railway with their own locomotives and rolling stock or arrange a working agreement with the GNR. The highly costly first proposal was quickly abandoned in favour of the latter and a letter was duly sent to King's Cross asking if the GNR would continue working the Holme to Ramsey line for 60 per cent of the gross receipts for a minimum period of two years.

The GNR Board of Directors examined the proposals on 22nd July, 1870 but the terms and period of time were considered inadequate and were curtly declined. The Directors, however, requested Henry Oakley, the General Manager, to negotiate with his GER counterpart Samuel Swarbrickand after several meetings spreading over the next two months, Oakley wrote on 30th September, 1870:

> The Great Northern is willing to work the Holme and Ramsey Railway at cost price, on condition that the proposed charges for working expenses shall not exceed 70 per cent of gross receipts and if traffic increases, not less than 50 per cent of gross receipts. The Great Northern will work the line for a period of five or seven years.

The tetchy relationship between the GNR and GER can be gauged by Oakley's conclusion:

> You will of course understand that I address you on this occasion as a representative of the Holme and Ramsey Railway and not in your post as General Manager of the Great Eastern Railway.

The terms were considered and agreed by the GER Board on 27th October. The GER General Manager advised Oakley that the five year term had been approved by his Directors. He also resolved to improve relationships and concluded,

There is no desire or intention on our part to alter the relative position of the company by reason of the negotiation for the working of the Holme and Ramsey Railway. At some point my Directors are desirous of putting an end to the anomalous position of that line and they trust your Board will be willing to consider the question with a view to an amicable settlement.

On 24th November, 1870 the GER effected the transfer of the Ramsey Railway stock between Directors of the company. The £17,900 ordinary shares and £10,000 debentures held in the name of James Goodson were transferred to the new Chairman Lightly Simpson, whilst the remaining £10,380 of shares were transferred from the holding of Colonel Jarvis to be shared equally between Swarbrick and Colonel William Thomas Makins and held in trust.

Edward Fellowes wrote to the GER from Haverland Hall, Norwich on 7th January, 1871,

> I have heard from Serjeant, Secretary of the Holme and Ramsey Railway Company that the Directors of the Great Eastern Railway desire that the money now in the hands of the Holme and Ramsey Directors be paid to the Great Eastern Company. You are aware at the present time that the two Directors of the Holme and Ramsey Railway, Mr Stafforth and myself each hold fifty shares. By the terms of the agreement entered into between the Great Eastern and the Holme and Ramsey Company when the Great Eastern bought the Holme and Ramsey line, the Great Eastern were to take all shares or pay five per cent on those they did not take. At the request of the Great Eastern, Mr Stafforth and I continue to hold our shares and act as Directors. Mr Stafforth is now a very old man and is anxious for the Great Eastern to take his proportion of the shares. We hold quite enough money to repay ourselves £500 each if the Great Eastern Directors so wish. If, however, the Great Eastern wish, I am willing to keep my shares and act as a director of the company. Mr Stafforth must be paid and I will pay the balance to the Great Eastern. [Fellowes concluded] It might be more advantageous to take all the shares and have the receipt money paid direct to the Great Eastern by the Great Northern Company and save the expenses of a secretary.

The letter was considered at the Board meeting on 19th January and 12 days later the GER Secretary advised Fellowes to transfer the shares held by Mr Stafforth to the joint ownership of Swarbrick and Colonel Makins. The Chairman was to retain his own shareholding and reduce expenses. No reply was received from Ramsey for two months and in the meantime the aged Mr Stafforth died. At the end of March 1871 his executor wrote to Bishopsgate asking the GER to pay £500 for the shareholding, adding that he was in no hurry for the money, provided the GER in the meantime paid five per cent interest.

A month later Fellowes contacted the GER and forwarded a cheque for £2,000, the Ramsey Railway portion of receipts belatedly received from the GNR authorities. By 6th July, 1871 Stafforth's executor advised that the transfer of shares to Messrs Makins and Swarbrick was to be finalized on payment of the necessary £500. Three months elapsed, however, before the transaction was finally completed on 12th October. A further £523 0s. 1d. was also paid to the GER as the outstanding portion of the Ramsey Railway receipts to the close of the financial year.

Meanwhile Henry Oakley, the GNR General Manager had written to his GER counterpart, requesting the extension of the siding at St Mary's to cater for increasing traffic. The cost was estimated at £230 and on 18th July, 1871, Davis the GER Engineer was asked to contact Johnson, his GNR counterpart regarding costs. After due deliberation it was agreed the sidings at both St Mary's and Ramsey were to be extended, and on 1st August the GNR agreed to transport the materials to site free of cost.

The death of Stafforth and transfer of his shareholding to the custody of the GER Directors soon created problems. Early in February 1872 Frederick R. Serjeant, the Ramsey Railway Secretary, advised that difficulty was being experienced in clearing cheques signed solely by Fellowes. Under the Ramsey Railway Act of 1861 the quorum of a minimum of two Directors was decreed. It was thought another Director should be appointed or failing that, the GER should authorize Fellowes to sign all cheques. Fellowes reported financial problems again in April 1873 when he advised the GER Board that a large amount of money was held in his name at the bank, but being the only Director of the Ramsey Railway he had no legal power of paying it over to the GER. He was frustrated with his role as the sole advocate of the Ramsey Railway and suggested that his shares be transferred to the main line company with effect from 31st March, together with all money outstanding, as soon as the March traffic receipts were paid over by the GNR.

If the proposal was adopted it effectively meant the Holme and Ramsey Railway would be totally absorbed into the GER. Such a move necessitated an Act of Parliament, and on 7th May, 1873 the GER Legal and Parliamentary Committee was requested to make application to Parliament in the next session, to authorize the GER to take the complete shareholding of the smaller company, paying in return the sum equivalent to their initial value. Fellowes was advised of the proposal but was becoming impatient and again wrote asking to be relieved of his shares because of the increasing problems they brought. On 21st May, Lord Claud Hamilton was asked to negotiate personally with Fellowes and by 18th June he had persuaded the Member of Parliament to wait until the Parliamentary affairs were completed. In the meantime it was arranged to place in deposit the money held, to enable the GER to obtain the advantage of an increased rate of interest.

With all attention focussed on the negotiations between the Ramsey Railway and the GER, the GNR as the operating company, continued to improve the train service including connections with main line trains at Holme. Considerable time was lost, however, at the junction when vans required unloading or when passengers requiring the Ramsey branch were located in the wrong portion of the main line train. To obviate delays and to increase accommodation for the Ramsey branch train, authority was given on 17th October, 1873 for the platforms at Holme to be lengthened at a cost of £200. The work, which considerably improved the smoothness of operation, was completed the following June.

Relationships between the GNR and GER were also improving and after several meetings the Directors of both companies announced on 8th May, 1874, their agreement on the Ramsey Railway and other problem areas where respective territories converged. The terms of the agreement to be included in the Bill submitted to Parliament were:

1. The Ramsey Railway to be leased to the GNR for 21 years from 1st July, 1875, on payment of two per cent upon an agreed capital of £43,000, such dividend to increase at the rate of a quarter per cent per annum each year until the maximum dividend of 3¾ per cent was reached. The arrangement was to include provision in respect of any expenditure incurred on new works. In the event of further works being required on the Ramsey line for the accommodation of traffic, the GNR was to bear the expenses subject to the approval of the GER - the GER repaying the GNR for such outlay at fair valuation for the works executed at the time of the renewal or the termination of the lease, such works to be treated as part of a going concern and not as mere land and materials only.
2. In return for services rendered, the GER to grant the GNR running powers from Victoria Park Junction to Victoria Docks and North Woolwich, on the same terms granted to the London & North Western Railway under the agreement of 10th April, 1872.
3. The GNR to undertake the maintenance of the GER line from the junction with the GNR at Shepreth to the GER junction at Shelford (Shepreth Branch Junction), for a period of 21 years from 1st July, 1874. The maintenance to include permanent way, ballasting, switches, crossings, sidings, gates, fences, quicks, ditches, signals and signal boxes, but not including stations or other buildings, nor the junction points. For this work the GER to allow out of the tolls payable by the GNR for the use of that portion of the GER, £320 per mile per annum.

Samuel Swarbrick and Henry Oakley duly signed the draft agreement on behalf of the GER and GNR respectively on 20th May, 1874.

By October negotiations were sufficiently advanced to the position that Edward Fellowes announced his proposal to withdraw the GER share money out of his bank. After deducting the £500 for his shares and discharging any outstanding debt, the balance was sent to the GER headquarters and registered as such by the GER Board on the 21st of the month. On 15th December, 1874 the agreement between the GER and GNR regarding the takeover of the Ramsey Railway by the GER and the subsequent lease to the GNR, with other working arrangements was duly signed and sealed by J.B. Skeggs and W.H. Shaw, Solicitors on behalf of the GER and Lord Colville and Reginald Capel, Directors of the GNR. The proposition was subsequently placed in the Private Bill Office of the House of Commons on 23rd December, 1874, for inclusion in the GER General Powers Act.

The Bill duly received the Royal Assent on 19th July, 1875, as part of the GER General Powers Act 1875 (38 and 39 Vict. cap. cxxxiv). The statute amongst other things confirmed that the Ramsey Railway was dissolved and the GER granted and the GNR accepted a lease of the line for 21 years from 1st July, 1875, under terms agreed between the companies on 15th December, 1874. The terms of the lease authorized the GNR to pay the GER rent calculated at two per cent per annum on £43,000, the original cost of the line. The rate of interest was to increase twice yearly on 31st December and 30th June at the rate of one quarter per cent to a maximum of 3¾ per cent per annum. The GNR was to bear the cost of any work on the railway subject to the approval of the GER and after 21 years the GER would repay the GNR a fair valuation of the works executed.

Holme station facing south *circa* 1875, with its low platforms and ballast covered permanent way. The up line is on the left and down line on the right. The Ramsey branch trains usually departed from the back of the island platform with its miniature waiting shelter. Note the water column beyond the level crossing, serving both the Ramsey branch and the up main line and the absence of a footbridge. The old style slotted signal is also of interest. The station master's house is at the south end of the down platform with booking office and waiting rooms alongside, whilst the nameboard proclaims 'Holme, change for Ramsey'.

British Rail

Chapter Three

Pre-Grouping Days

Frustrated by the vain attempts over the years to extend the railway east of Ramsey, local businessmen finally resurrected proposals for a railway linking the Ramsey Railway with the St Ives to March line at Somersham. The Bill was duly presented on 9th February, 1875, and subsequently received the Royal Assent on 13th August, 1875 as the Ramsey and Somersham Junction Railway Act 1875 (38 & 39 Vict. cap. ccxii). The Act authorized the company to construct a line, 7 miles 3 chains or thereabouts in length, situated wholly in the County of Huntingdon, commencing in the parish of Ramsey by a junction with the Holme and Ramsey Railway, 8 chains or thereabouts west of the booking office at Ramsey station. The line was to terminate in the parish of Somersham by a junction with the St Ives to March line of the GER, 14 chains or thereabouts north of the booking office of Somersham station. Three years were allowed for the compulsory purchase of land and five years for the completion of works. The cost of construction was to be raised by the sale of 5,000 £10 shares with powers to borrow £16,666, once £25,000 had been paid up. (For the full history of this line see the companion volume *The Ramsey East Branch*.) Both the Ramsey Railway and the new Ramsey & Somersham Junction Railway (R&SJR) were closely tied, as Frederick Serjeant, the Ramsey solicitor, was Secretary to both companies.

The final payment of the proportion of receipts received from the GNR by the Ramsey company was paid over to the GER on 22nd September, 1875, when Edward Fellowes forwarded a cheque for £3,037 7s. 6d. to conclude the affairs of his former company.

At this time the GER was desperately trying to infiltrate the area north of March to obtain access to the northern counties and the lucrative coal traffic. The GNR favoured an amalgamation of routes on equitable terms but the GER fought to build an independent line fearing any amalgamation would give the GNR access to East Anglia. In a report presented on 5th September, 1877, from Sir John Hawkshaw regarding possible routes, two of the possible lines suggested south of Lincoln involved the construction of a railway from Long Stanton to Somersham, Warboys and Ramsey. The first ran thence to Peterborough and Bourne, whilst the second ran from Ramsey via Whittlesea and on to Thorney. The proposed lines would make junctions with the GER at Somersham and Whittlesea but no mention was made of a junction with the existing branch at Ramsey. The R&SJR was the subject of discussion at the GER Northern Extension Committee meeting on 25th January, 1878.

Despite the initial euphoria regarding the through line from Holme to Somersham, the R&SJ made no progress in the three years allowed by their Act for the compulsory purchase of land. In May 1878 the Town Clerk of St Ives reported that there were rumours that the GER and GNR had abandoned plans to construct the extension. He reported that his council considered the extension essential, as much of the trade passing between St Ives, Ramsey and

Peterborough was conveyed throughout by water transport. A copy of the letter was passed to Henry Oakley of the GNR. He duly wrote to his GER counterpart Samuel Swarbrick, 'The extension had never been considered as a joint exercise' and asked what the present position was. He reminded him that such local lines were 'very costly to work' and thought a tramway would answer the purpose. The Directors visited St Ives as part of a tour of inspection of the line on 18th May and raised the matter at their Board meeting four days later, when it was decided to defer the subject for the time being. The promoters were subsequently forced to petition Parliament for an extension of time. On 4th July, 1878 the Ramsey and Somersham Junction Railway Act (41 & 42 Vict. cap. cxliii) was passed and duly authorized the company to extend the time allowed for purchase of land by two years and completion of works to three years from 13th August, 1878.

Having obtained a long lease on the line, the GNR settled down to operate the Ramsey branch services but the calm of this railway backwater was shattered by a tragic accident, which occurred at Holme station. On 30th September, 1878, as a group of schoolchildren were crossing the main line by the wicket gate crossing, an express approaching at speed bore down on the small company. Most scattered in terror but unfortunately Alfred Rhodes, a five-year-old was run down and killed. An immediate investigation was called and on 25th October the GNR Engineer reported to the Directors that the children from the village regularly crossed the main line and branch line level crossings to get to and from school. The coroner, whilst recording a verdict of accidental death, was highly critical of the railway company officials and recommended protection for the children by the appointment of a gateman to attend to the gates when the children were passing to and from school. The Engineer recommended the provision of a footbridge to save the cost of a gatekeeper but whilst noting the idea, the Directors requested a daily census of people using the level crossing. At the GNR Board meeting on 13th December, 1878 the Engineer reported that on average 250 adults and 28 children crossed the railway daily. Evidently the Directors were concerned by the totals for authority was duly granted for a footbridge to be erected adjacent to the level crossing at Holme, the work being completed in the spring of 1879. The bridge was of a recently developed suspension construction, with a span of 80 ft, known as Harper's Patent Foot Passenger Bridge. The *Peterborough Advertiser* for 26th April, 1879 gave a full description of the structure:

> On Monday last, a new suspension bridge (known as Harper's Patent Foot Passenger Bridge), placed across the Great Northern level crossing at Holme Railway Station, was opened to the public.
>
> The bridge, which is of recent invention, is said to possess the combined principles of suspension, tension and arch, and was originally designed at the request of several owners of extensive coffee plantations in Ceylon, Africa, and elsewhere, for the purpose of spanning rivers, ravines, &c. It has also been adopted by several noblemen and gentlemen in Scotland, one erected on the Marquis of Huntly's estate at Aboyne, crosses the River Dee with a span of 300 feet.
>
> The one just opened has a span of 80 feet, with 6 feet clear width for traffic, and is light and pretty in appearance: a decided improvement upon the unsightly wooden

structures erected for the same purpose over some of our railways. As proof of its portability and simplicity of construction we may state that it was erected in seven days', without in any way interfering with the traffic on the line, which at that point is very considerable, upwards of 1,300 trains (including shuntings) passing under whilst in the course of construction. The work was executed under the personal supervision of Mr John Harper of Seafield House, Aberdeen, the patentee.

At this time residents of Ramsey frequently complained of the poor facilities for passengers at the station but in spite of regular correspondence with the railway company no improvements were made. Meeting with no success local townsfolk requested Edward Fellowes, in his capacity as former Chairman of the Ramsey Railway, to intervene. Fellowes wrote to the General Managers of both the GER and GNR asking if the present wooden building, which was considered to be a temporary structure, could be replaced by a more permanent building. The GER on receipt of the correspondence immediately asked if the GNR authorities were prepared to improve the wooden station building or alternatively erect a new one. After some discussion, the GNR advised both Fellowes and the GER that only minor improvements would be made to the existing timber structure, as 'passenger receipts from the line barely warranted the expense of a new station!'

The GER attempt to forge a link with railway companies in the North of England and the GNR pressure to gain access to East Anglia reached impasse after further confrontation in 1878, when the GNR obtained powers to build a Spalding to Lincoln extension and the GER Northern Extension plans were rejected. Whilst authorizing the GNR new works Parliament added a rider to the effect that the GER ought to have access to the North. The possible costly duplication of lines was thus fortunately avoided when the GNR dropped its demands for running powers to Norwich, Yarmouth and Lowestoft, and proposed a joint line to the North commencing at Huntingdon. The GER authorities agreed to the proposals and on 3rd July, 1879 the Great Northern & Great Eastern Joint Committee (GN&GE) was authorized, the Board formed of five Directors of each company. The new joint line ran on existing lines from Huntingdon to St Ives, with running powers thence over the GER to Needingworth Junction, then from Needingworth Junction via Somersham to March South Junction, March Whitemoor Junction to Spalding, the authorized Spalding to Lincoln section, completed in August 1882, and then via GNR lines to Gainsborough and on to Black Carr Junction just south of Doncaster where a connection was made with the GNR main line.

In 1880 the R&SJ Directors were concerned that the time permitted for the purchase of land for their railway granted in the 1878 had almost expired. Parkes of the GER had initially offered a £25,000 subscription to help finance the railway, as well as work the line at cost price, but as the GNR showed complete indifference to the scheme the offer was withdrawn. Faced with little or no alternative source of financial backing, the R&SJ Directors, early in July 1880, reminded the GER Board of their promise and moral obligation to uphold the offer. During the ensuing debate at the GER Board meeting on 18th August, it was thought that the Joint Line Northern Extension scheme would make the new line unnecessary, but the Directors again suggested that if the offer of the subscription were taken up, it should be jointly owned, half by the GER and half by the GNR. If this was acceptable, the Ramsey to Somersham line would be jointly owned, and in return the GER would sell half of the shares of the Holme to Ramsey line to the GNR. The Ramsey and Somersham promoters fully concurred and at the following GER Board meeting on 14th September, 1880, the Directors agreed to suggest to the GNR that they purchase half of the Holme to Ramsey railway and the extension be considered by the GN&GE Line Joint Committee.

However, by now the GNR was totally disinterested in extending their territorial rights into East Anglia via Ramsey, and the hope of providing a through route to Somersham was again thwarted. The R&SJR thus returned to Parliament seeking a further extension period. The Royal Assent was duly given on 8th April, 1881 as the Ramsey and Somersham Junction Railway Act 1881 (44 and 45 Vict. cap. xii) granting powers to extend the time limited for compulsory purchase of land to two years from 13th August, 1881, and three years for the completion of works from the same date.

Further efforts were made by the R&SJR Directors to engender support from the main line companies but each application drew negative responses. The GER was adamant they would not finance the company unless the GNR provided half of the cost. The GNR dogmatically opposed any such idea as they had no wish to draw traffic from the existing branch from Holme, which to all intents and purposes provided only a meagre revenue compared to other branches on their system. On 5th December, 1882 at the GER Board meeting the R&SJR was again a topic of discussion. C.H. Parkes (Chairman of the GER) advised that the matter had been discussed by the GN&GE Joint Committee and that in view of the local support for the line, either the GER or GNR or the Joint Committee should work the railway when completed, with the GER taking over the working of the Ramsey to Holme line if the GNR wished to part with it. Oakley was adamant that the GNR Directors saw no economic prospects for such a venture and after a further meeting between the two General Managers the position was stalemate. Thus after two years of unsuccessful manoeuvring no financial backing was forthcoming and the R&SJR Directors were again forced to return to Parliament. Authority was subsequently given in the Ramsey and Somersham Junction Railway Act 1883 (46 and 47 Vict. cap. cl), which received the Royal Assent on 2nd August, 1883, for extension of time to purchase land for two years from 13th August, 1883, and three years for the completion of works.

On the afternoon of Wednesday 31st October, 1883, John Julyan, a 67-year-old farmer from Bury, was crossing the line at Ramsey station when he was struck by the branch engine as it ran round the train, receiving severe injuries to his left leg. A passenger and Mr Atter, a friend of the farmer, witnessed the accident, as did the engine driver who immediately stopped the locomotive. The engine was immediately re-coupled to the train, which formed a 'special' and conveyed Julyan to Peterborough, from whence he was taken to the infirmary. On arrival it was found his left leg was crushed below the knee and his right foot was mangled. As he was suffering from shock, no amputation was made until the following day when his condition had improved. The left leg was amputated above the knee and the right leg above the ankle but by Friday, Julyan's condition deteriorated and he passed away. At the inquest held at Peterborough on Saturday afternoon, the jury merely viewed the body before reconvening on Monday 5th November. In evidence it was stated when the engine had reversed, the driver did not sound the whistle, and Mr Gachee, representing the railway company, stated it was 'not the usual practice to do so'. He added there was no 'right of way' where the deceased crossed the line at Ramsey. The jury subsequently returned a verdict of accidental death.

Meanwhile the flow of agricultural traffic handled by the GNR at Ramsey had increased to such an extent that by late 1883 the cramped layout was causing operating problems. The local goods manager requested the urgent installation of additional siding accommodation to relieve congestion. The GNR authorities at King's Cross were still dubious of making a large outlay on a branch line owned by the GER, until they had negotiated with Bishopsgate. The request for the new siding, with dock, wharf and cattle pens costing £369 7s. 6d., the enlargement of the goods office costing £117 15s. 6d. and alterations to the passenger station at a cost of £212 11s. 0d. were considered and agreed by the GER Board at their meeting on 1st January, 1884. Evidently the estimate for the provision of the siding was too low for on 7th March, 1884, the authority was increased, provided the outlay did not exceed £500. Later in the year an additional siding was required at St Mary's to handle root crop traffic and on 3rd October, 1884, the expenditure of £111 was sanctioned for the new works. The sidings at Ramsey were completed in time for the harvest in August 1884 but the additional accommodation at St Mary's was not completed until late March the following year.

By 1885 the R&SJR Directors had exhausted all hopes of monetary assistance from either the GER or GNR companies and decided to raise the capital for their line unaided. A consortium of three purchased all the shares so that work could commence, but despite the authority granted by the original Act, the railway terminated half a mile distant across the town from the existing Ramsey station. Work on the Somersham line was completed by August 1888 and the following month Major F.A. Marindin, the Board of Trade inspector, refused the opening of the line to passenger traffic because of the incompleteness of works. The failure of the new line to connect with the Holme to Ramsey branch brought the following comment at the conclusion of his report.

> There is no junction from the new line to the existing line to Holme on the Great Northern Railway, this small country town of Ramsey has therefore two terminal stations half a mile apart, which is a great public inconvenience. It seems to be that such a thing has been sanctioned simply from the opposition of one railway to another but as they have thought fit to do so, I presume the Board of Trade has no powers in the matter.

In the meantime the GN&GE Joint Committee had discussed the possible purchase of the R&SJR. Birt and Oakley, the respective General Managers of the GER and GNR wanted to know if the deal should include the Ramsey Railway. The GER Board on 1st November, 1887 asked the General Managers to convene another meeting before again approaching the Committee. No further action was taken and in March correspondence was awaited from the Somersham company.

On 17th January, 1889, Oakley of the GNR and Birt of the GER, as respective General Managers were requested to negotiate a working agreement with the R&SJR. On 4th March it was agreed the GER would work the new line, initially for a period of five years, for 70 per cent of the gross receipts. Clause 3 of the agreement established that the GER would compete in the interests of the Somersham company in competition with the GNR in the same mode as the companies vied with each other at Spalding, whilst clause 7 stressed that any loss to the GNR by reason of the GER opening to Ramsey and carrying the Ramsey company traffic would be contributed to by the GER out of any amount the GER would derive from such traffic, after deducting the terminal and working expenses at the rate of 50 per cent, such losses being ascertained annually.

After some remedial work and the agreement of the working arrangements by the BoT, the R&SJR opened to traffic on Monday 16th September, 1889, 14 long years after the line was originally mooted and 24 years after the original plan by the GER, but leaving what could have been a useful through route, as two almost non-descript branch lines. On the opening of the new branch the terminal station was named Ramsey High Street whilst the existing station retained the name Ramsey.

The Regulation of Railways Act 1889, amongst other things, required the interlocking of points and signals on all lines and the installation of the block telegraph on all lines, except those worked under the Train Staff without Tickets or 'One Train Only' conditions. The GNR responsible for renewing equipment on the branch under the leasing agreement, duly advised the GER that it had more important lines to bring up to standard before the Ramsey branch. The GER in a similar position, agreed that the BoT be requested to sanction postponement of the work. After correspondence between the parties, the BoT agreed to grant an extension of time to 20th November, 1892, to complete the interlocking of signals and points on the Holme to Ramsey line.

Despite the provision of the additional siding at St Mary's in 1885, the goods handled at the station continued to increase so that the limited facilities provided were unable to cope and wagons had to be detained at Peterborough, Holme or Ramsey until farmers, growers and coal merchants had offloaded wagons. Several complaints were made by local growers regarding the delay in transit of their produce to markets. To ease the situation the GNR authorities sanctioned the provision of an additional siding in the goods yard at an estimated cost of £200 on 7th November, 1890. The siding was duly installed in May 1891.

By powers authorized in the Great Northern Railway Act 1891 (54 Vict. cap. xix), which received the Royal Assent on 11th May, 1891, the company was authorized to widen their main line between Abbot's Ripton and Fletton. Five years were permitted to complete the works, which included the widening of level crossings over public roads Nos. 14, 41 and 44 in the parish of Holme. Clause 10 of the statute required the company to lay only one line of rails over the former, unless an existing siding, which ran across the road was removed. Because the widening affected the junction with the Ramsey branch, clause 18 sub-sections (1) and (2) stipulated the GNR was not to interfere with any junctions or connections at Holme, or any lands or works owned by the GER, or commence any works until drawings of the proposals had been presented and approved by the GER Engineer. Sub-section (3) stipulated that during construction the GNR was not to obstruct or interfere with traffic passing along the GER Holme to Ramsey branch, whilst sub-section (4) granted the GER similar rights and powers over the altered junction as enjoyed over the existing junction. In the event of any lasting dispute either party could call upon the President of the Institute of Civil Engineers to appoint an arbiter to settle the differences.

During a dry spell of weather in July 1891 a locomotive hauling the branch mixed train succeeded in belching forth sparks which set fire to crops and farm buildings owned by J. Newton. As a result of the incident animals were injured. A claim was subsequently lodged against the GNR for £1,377, when it was alleged the locomotive was improperly constructed and managed, thereby causing sparks to set fire to farm property. The action was contested but so grave was the allegation the case continued well into the next year. On 27th May, 1892, the GNR solicitor advised his Directors that the services of the Attorney General were being sought and application was being made for the case to be heard in London instead of Huntingdon.

The General Manager advised the GER Board meeting on 6th June, 1893 of the working of the R&SJR line in which the GER received 65 per cent of earnings up to £10 per mile. In return the GER compensated the GNR for traffic abstracted from the Holme to Ramsey line and the GNR compensated the GER for any loss incurred working the line. From its opening on 16th September, 1889 to 1st June, 1892 it was estimated the GER owed the GNR £1,536 for the abstraction of traffic and the GNR owed the GER £3,785 as a share of the loss.*

On 15th November, 1893, a serious incident occurred at Ramsey station when a light engine running tender first from Holme to Ramsey, to be attached to the 5.55 pm up train from Ramsey to Holme, entered the station at excessive speed and collided with four coaches standing at the platform. As a result of the impact the coaches were driven back for a distance of 10 to 12 yards but none became derailed. Sixteen passengers had already boarded the train and ten of them complained of injury, whilst the brakesman riding on the engine bruised his shoulder. The engine and tender were not damaged but the dome lights of three brake carriages in the train and the buffer beam and castings of the brake carriage struck by the tender were broken. On 27th November the BoT ordered an inquiry into the accident and the task was designated to Major F.A. Marindin.

In his investigations Marindin found that Ramsey station, with a single platform on the south side of the railway was the terminus of a single line 5 miles 58½ chains in length from the junction at Holme. Approaching the station there were three facing point connections from the main single line leading to the engine shed, goods yard and sidings respectively. The outermost set of points was 228 yards west of the point of collision, which was about five yards east of the outer end of the single platform. The facing points were normally locked for the main single line and were worked from a ground frame, whilst the down home signal and up starting signal were mounted on the same post

---

* There is a further explanation on page 43.

located at the west end of the platform. The only other signal was the down distant, located 543 yards west of the home signal. The signals and points were not interlocked and the line was worked on the 'One Engine in Steam' principle with the driver carrying the single line Train Staff.

Giving evidence, driver James Sutton stated he had been a railway servant for nearly 19 years and a driver for four years. He was well acquainted with the Ramsey line having frequently driven across the branch during the previous three years and as a fireman for a number of years before that. On the day of the incident he came on duty at 12.25 pm to work until shortly after 10.00 pm. He had six-coupled tender locomotive No. 340 running tender first from Holme to Ramsey. The first train he worked was the 1.14 pm from Holme, culminating in the 4.30 pm train from Holme arriving at Ramsey at 4.44 pm. The next trip was the 4.50 pm up goods from Ramsey, which was due to arrive at Holme at 5.27 pm. The train departed to time but because of the considerable shunting required at St Mary's the train did not arrive at Holme until 5.40 pm. The train was shunted into the siding and the engine set off light back to Ramsey at 5.49 pm, in order to work the 5.55 pm up passenger train from Ramsey. Sutton advised the inspector that he had driven the engine at normal speed, noting the signals at St Mary's were clear. Running at some 30 mph he saw Ramsey distant signal was at danger at a distance of a quarter of a mile, shut off steam and reduced the speed of the engine to approximately 20 mph by the time he passed the signal. The vacuum brake was slightly applied twice between the distant signal and the level crossing located 250 yards from the point of collision. By then Sutton estimated the engine was travelling at 15 mph and the brake was then fully applied. Unfortunately because of the drizzly rain and the greasy condition of the rails, by the time the engine passed the engine shed, 103 yards west of the point of collision, the driver realised the locomotive would not stop and he reversed the engine and applied steam to prevent the collision. To further questioning, Sutton advised Marindin that he estimated speed at the time of impact was 4 mph. He remained on the footplate at all times and was not knocked down or injured as a result of the collision. He admitted he knew the coaches were at the platform and that he ought to have made a more steady approach to the vehicles. Under cross-examination, the driver stated that all six wheels on the engine and six wheels on the tender were braked. Although the automatic vacuum gauge showed a vacuum of 20 inches, he could not remember what the reading had reduced to when he applied the brake at the crossing. He did not think it necessary to apply the brake more fully at the crossing and estimated the engine was travelling at 12 mph when passing the engine shed. The light drizzle had come on after the light engine departed from Holme, which made the rails very greasy. He thought he had made due allowance for this and was confident the engine was going to stop. Sutton reported the fireman had screwed down the brake on the tender and some of the wheels were skidding at the time of impact. The fireman also applied the sand soon after passing the crossing but because the engine was running tender first, the sand would have only acted on the leading two wheels of the locomotive. The brakesman, who was riding on the engine called out after he had fully applied the brakes to 'steady the engine', but there was nothing he could do to prevent the collision. Because of the situation, Sutton admitted he had not sounded the engine whistle to attract attention.

The next to give evidence was William Brown, the fireman who stated he had been 16 years in railway service and a fireman for 12 years. He was well acquainted with the Ramsey branch and concurred with the driver's statement, except that he did not apply the handbrake at all. He recollected the engine was travelling between the crossing and the shed when he opened the sand valves but could not say exactly where the driver applied the brakes as he was attending to the injector. Brown concluding by telling Marindin he did not feel alarmed at the speed the engine was travelling and thought they would stop before reaching the coaches. He was knocked over by the collision but was uninjured.

The inspecting officer then called Charles Edward Robinson to give evidence. He advised Marindin he had been in the service of the GNR for two years and seven months and a brakesman for 13 months, all of which had been spent on the Ramsey branch. On 15th November he had signed on duty at 12.25 pm to work through to 11.10 pm. He reiterated the driver's evidence and reported the light engine had left Holme on the return journey to Ramsey 12 minutes late at 5.49 pm. He was of the opinion speed on the return journey was not much above average. The journey was made in 12 minutes, one minute less than scheduled. Robinson advised that the driver had applied the brakes slightly and reduced the speed of the engine on passing the distant signal, which he confirmed was at danger. Although Sutton had made two further brake applications before reaching the crossing, the brakesman thought the speed of 10 to 12 mph was too high and he called the driver to reduce speed further and for the fireman to apply the handbrake. He confirmed the engine was thrown into reverse and speed at the time of the collision was about 4 mph. The brakesman was thrown against the end of the tender and slightly bruised his shoulder. To further questioning Robinson advised the inspector that the time of the accident was 6.01 pm. The train at the platform consisted of four carriages, one brake/third, one brake composite, one composite and another brake/third.

Last to give evidence was station master George Vincent who had been in charge at Ramsey for four years out of a total service of 27 years with the GNR. He confirmed he had been on duty on 15th November when the collision occurred and that the coaches were standing at the platform fully prepared to depart, once the engine arrived from Holme and coupled up. There were 16 passengers on the train. Vincent told Marindin he saw the engine approaching and it seemed to be travelling faster than usual past the engine shed. He was not alarmed and thought the driver would stop the locomotive in time so no attempt was made to get the passengers out of the train. No whistling was heard and the speed at the time of impact was about 4 mph, pushing the train back about one carriage length. He immediately went to the assistance of the passengers and received complaints of injury from five persons. He thought driver Sutton was a 'steady man' and after the accident asked how he could account for the collision, to which the driver had admitted he had 'miscalculated his speed a little.' Vincent booked the time of collision at 6.00 pm but the guard (brakesman) had registered it at 6.01 pm, attributing the difference to the variation between the brakeman's watch and the station clock.

In his report Marindin placed the blame for the collision fairly and squarely on driver Sutton for failing to get his engine under proper control when approaching Ramsey station. Running some 13 minutes late, the 5 miles 58 chain run had been made in 12 minutes, one minute less than the booked time. The inspector concluded the driver had been 'hurrying back' and that as the proper time for departure of the up passenger train was already past, he miscalculated the speed and failed to make sufficient allowance for the greasy condition of the rails, allowing his engine to run too far before making use of the brake acting on all the wheels of the engine and tender. The engine overran the home signal and came into collision at a speed of 4 mph with the loaded carriages of the train waiting at the platform.

As a result of the accident, the BoT inspector severely criticized the GNR for persistently working the Holme to Ramsey line with tender locomotives instead of a tank engine.

> It is much to be desired that, on branches where there are no facilities for turning engines, properly constructed tank engines should be used for working passenger trains, for the running of engines tender first involves a certain amount of risk, and exposes the drivers and firemen to great discomfort in bad weather. It is more than possible that if the engine in this case had been running in front of the tender, or if it had been a tank engine, the accident would not have occurred, for in either of these cases the sand applied to the rails would have affected all the wheels, instead of only two pairs of the engine wheels.

The GER did not emerge unscathed, for the inspecting officer reminded the company that under the powers of the Regulation of Railways Act 1889, the points and signals on the Ramsey line should have been interlocked before the deadline of 20th November, 1892 but no work had been carried out. Marindin concluded, 'It is desirable that the GER Company should be called upon to cause the necessary work to be carried out at once by whichever company may, by terms of the lease, be responsible for the cost thereof'.

The GER General Manager (William Birt) reported to his Board on 6th February, 1894 that the actual earnings of the R&SJR were about £6 per mile per week and the GNR share of the loss from the date of opening to 30th June, 1893 was £9,658. The GNR General Manager after consulting with his Directors advised Birt to refuse the renewal of the working agreement and gauge the response from the R&SJR Directors.

In the meantime copies of Marindin's report on the accident had been passed to the GNR and GER on 1st January, 1894, and the BoT subsequently approached the GER as the owning company and the GNR as the leasing company, to comply with the requirements of the Act as a matter of urgency. It was 20th February, 1894 before the subject was raised at the GER Traffic Committee, when Birt the GER General Manager reported that the 5 miles 65 chains of line was the property of the GER but leased to the GNR for a period of 21 years from 1st July, 1875. Johnson of the GNR estimated the cost of new signalling and interlocking of points and signals at £2,008, which was considered a fair price by the GER signal engineer. Birt advised that under the leasing agreement the GNR would pay for the work and reclaim the capital at the termination of the agreement. The Traffic Committee readily approved of the outlay and the contract was awarded to McKenzie & Holland.

The subject of the R&SJR and the Holme to Ramsey line was again the subject of detailed discussion at the GER Board meeting on 2nd October, 1894. Birt explained that the two lines were closely intertwined as the GER worked the R&SJR for 65 per cent of the earnings and compensated the GNR for any abstraction of earnings from the Holme branch, which the GNR rented from the GER at £1,612 10s. 0d. per annum and £2,648 for the year ending 30th June, 1894. The GNR in turn paid a share of the losses of the R&SJR, £5,306 out of a total of £10,612. The net value of the traffic between the GER and the R&SJR was £4,773 resulting in a net loss of £533. The R&SJR was 6 miles 6 furlongs in length and the Holme to Ramsey line 5¾ miles, the latter being bought by the GER at a cost of £43,000. The two stations at Ramsey were only half a mile apart but the expense of connecting them would cost more than any savings, which might be made. The net earnings of the Somersham to Ramsey line were a mere £780, £1 3s. 0d. per cent of the authorized capital of £66,666. As the GER agreement with the R&SJR and the GNR both expired on 30th June, 1894 the General Manager was instructed to discuss with his GNR counterpart to offer to purchase the R&SJR for no more than £30,000 and extend the existing agreement to the end of 1895.

The installation of the new signalling works was nearing completion on 22nd September, 1894, when the GNR Secretary advised the BoT that the work would be available for inspection within ten days. Major F.A. Marindin conducted the official BoT inspection of the new works on 26th October, 1894. He found that the points and signals at St Mary's and Ramsey were fully interlocked. At St Mary's a new raised signal box was provided with a 20-lever frame containing 14 working and six spare levers. One of the working levers bolt locked the gates of the adjacent level crossing, whilst an outlying 3-lever ground frame was provided for entrance points to the goods yard. This lever frame was locked by Annett's key, which was kept in the signal box except when required to operate the lever frame. When the ground frame was in use the signals on the main single line were locked in the danger position. At Ramsey the new signal box contained a 25-lever frame with 17 working and eight spare levers. Marindin also noted there was a 3-lever ground frame, bolt locked from the signal box, which controlled the points by the buffer stops at the end of the platform road. The inspector found the interlocking was correct at both stations and recommended use of the new signalling. He was somewhat puzzled by the fact that the branch, worked on the 'One Engine in Steam' method of operation was fully signalled, for he concluded his report pointing out that it might have been possible to dispense with such luxury, except at the level crossings, provided it suited the working of the traffic, to lock the points with a key forming part of the single line Train Staff. During the inspection Marindin, added that in his opinion the accommodation at both stations but especially at Ramsey, was 'very poor' and trusted that steps would be taken to improve the facilities. Marindin's report and recommendations were sent to King's Cross on 27th October and the

GNR Secretary, noting the criticism of the stations, promised improvements. When the BoT reminded the company in the spring of 1895, the Secretary replied that as the lease of the GER-owned line expired in July 1896, the GNR was 'not prepared to incur any expense improving the stations'.

The Great Northern Railway Act 1895 (58 Vict. cap. xxxvi) passed on 30th May, 1895 amongst many other things authorized the company to close the level crossing south of Holme station and divert the public road known as Long Drove on a bridge to be built across the line. Clause 29 permitted the extinguishing of all rights of way in and over the line as well as in and over the footbridge crossing the railway. The diversion of the road was to commence about 190 yards west of the level crossing and terminate 250 yards east thereof. The work was not, however, carried out. Separate authority was given for resignalling work at Holme where the existing signal box containing a 42-lever frame was totally inadequate for the expanding facilities at that location. As an expedient, temporary arrangements were made by adding an additional six levers to the existing frame pending provision of the new signal box. The contract for the new works was awarded to Saxby & Farmer.

The GNR had made regular use of the wharf alongside the Holme Navigation for transhipping produce and goods into and out of fen lighters, which provided a feeder to the railway from outlying parts of the fens. By the spring of 1895 the Holme Brook section was silting up and, on 7th June, authority was given for £185 to be spent dredging the portion of the Navigation to provide a deeper channel beside the wharf.

On 30th January, 1896, the GNR Secretary advised the BoT that resignalling works at Holme would be ready for inspection by 8th February. Lieutenant Colonel H.A. Yorke duly carried out the inspection on 8th March and found the signal box now contained a 65-lever frame, with 60 working and five spare levers. The interlocking between points and signals was correctly carried out. Whilst at the station, Yorke noted the waiting shelter on the up side island platform provided poor accommodation for passengers interchanging between the main line and the Ramsey branch. Of particular concern was the narrowness of the platform, especially on either side of the shelter. The inspector strongly recommended, if possible, the slewing of the branch line outward to enable the platform to be widened so that 'proper waiting accommodation and toilet facilities' could be provided. The GNR Secretary wrote to the BoT on 13th March agreeing to comply with the works.

Under the 1875 Act, the lease of the Ramsey branch by the GNR was due to expire on 30th June, 1896. As the arrangement between the GNR and GER was working satisfactorily no alteration was deemed necessary, save the authority of Parliament and the agreement was duly signed by T.D. Genlloud, the GER Secretary and F. Shuttleworth, his GNR counterpart on 17th January, 1896. The Northern company had duly arranged to include the application for renewal in a Bill presented to Parliament in December 1895, and this received the Royal Assent on 20th July, 1896 as the Great Northern Railway Act 1896 (59 and 60 Vict. cap. cxxxviii). Included in the statute was the authority for the GER to lease the Ramsey line to the GNR for a further period of 21 years from 1st July, 1896. The GER was to grant the lease at an annual rent of £1,612 10s. 0d., being 3¾ per cent of the £43,000 original capital of the Ramsey company. The GNR in return was required to work the line and maintain the works in good order. During the period of the initial lease the GNR advised the GER authorities that £2,996 2s. 6d. had been expended on the Ramsey branch and under negotiable terms the GNR received a proportion of this outlay in return for additional assets provided. The Act also transferred the Ramsey & Somersham Junction Railway to the GN&GE Joint Committee in confirmation of the agreement made between the two parties on 14th December, 1895.

Despite the promise made to the BoT on 13th March, 1896, no action was taken to improve the up platform at Holme and on 30th October, 1896 correspondence was received from the BoT asking what improvements had been made. The matter was placed before the GNR Board and on 7th November, the Secretary replied to the effect that the company was planning to widen the formation to accommodate four tracks through Holme station and would improve the platform as part of those works. Unfortunately the line was never widened and Holme up side platform was narrow until closure and subsequent demolition.

The importance of Holme as the interchange point for the Ramsey branch and the consequential growth of both passenger and freight traffic over the years deemed it necessary to recruit additional staff, especially for the freight function. Complaints were received from staff already employed at the station of the difficulty of finding suitable living accommodation in the area. The matter reached the level of General Manager in November 1896, who after raising the issue with the GNR Board on 16th of the month, received the authority to provide four cottages for staff near the station at an estimated cost of £850.

As a result of the revision of the operating practice on the branch, from the late 1880s a locomotive was no longer outbased at Ramsey and the engine booked to haul the services worked out and back to and from Peterborough. For some years the former engine shed at Ramsey built of corrugated iron on a timber framing, had been disused and was gathering cobwebs, birds nests and a colony of rats. By the winter of 1896 the building was in such a decrepit condition that the GN locomotive department advised the General Manager they wished to demolish the structure. Under the terms of the lease the GN advised the GER General Manager of the circumstances. The matter was raised with the GER Traffic Committee on 16th February, 1897 and, after learning that the building had been standing for 35 years and had an assessed value of only £50, they agreed to demolition. The GNR formally agreed to the sum being offset against the capital expenditure at the termination of the current lease in 1917.

Vegetable traffic continued to increase on the branch and during the root crop lifting season the facilities at St Mary's were again taxed beyond their limits. Often empty wagons were detained at Peterborough, Yaxley or Holme waiting until the branch goods train had cleared the yard. Produce was delayed reaching the markets and subsequent claims were made against the GNR for delivery of stale items. The situation was so desperate that by the autumn of 1897 the General Manager requested authority for the provision of a further

This Agreement made the 17th day of January 1896 **between** THE GREAT EASTERN RAILWAY COMPANY (hereinafter called "The Great Eastern Company") of the one part and THE GREAT NORTHERN RAILWAY COMPANY (hereinafter called "The Great Northern Company") of the other part

WHEREAS by an Agreement dated the 15th day of December 1874 and made between the Great Eastern Company of the one part and the Great Northern Company of the other part and which was sanctioned by The Great Eastern Railway Act 1875 the Great Eastern Company agreed to lease the Holme and Ramsey Railway (being the Railway which commences by a junction with the Great Northern Railway at the Holme Station of that Railway in the County of Huntingdon and terminates at Ramsey in the same County) to the Great Northern Company for the term of 21 years from the said 1st day of July 1875 on the terms therein mentioned AND WHEREAS since the 1st day of July 1875 the Great Northern Company have been in occupation of the Holme and Ramsey Railway as Tenants of the Great Eastern Company under the terms of the said Agreement but no lease of the said Railway has been granted AND WHEREAS since the 1st day of July 1875 some further works have been required on the Holme and Ramsey Railway and have been executed by and at the expense of the Great Northern Company under the provisions contained in Clause 3 of the said Agreement and the aggregate cost of the said further works according to the accounts furnished by the Great Northern Company amount to the sum of £2996 2s. 6d. AND WHEREAS the term agreed to be granted by the hereinbefore recited Agreement will expire on the 1st day of July 1896 NOW IT IS HEREBY AGREED AND DECLARED by and between the parties hereto as follows:

1. THE Great Eastern Company will grant to the Great Northern Company and the Great Northern Company will accept a lease of the Holme and Ramsey Railway for a further term of 21 years from the 1st day of July 1896 at the yearly rent of £1612 10s. 0d. (being 3¾ per cent. on the Capital of the Holme and Ramsey Railway amounting to £43,000) payable half-yearly on the 31st day of December and the 30th day of June in each year of the said term and under and subject to the covenants agreements and provisions hereinafter mentioned or referred to.

2. THE lease to be made in pursuance of this Agreement shall contain covenants on the part of the Great Northern Company

    (*a*) To pay the said yearly rent of £1612 10s. 0d. free from any deduction (except property tax thereon) by equal half-yearly payments on the days aforesaid.

    (*b*) To bear pay and discharge all rates taxes charges and impositions whether parliamentary parochial or otherwise which during the said term shall be charged assessed or imposed on the demised premises or on the landlord or tenant in respect thereof (excepting property tax on the said rent).

    (*c*) To uphold maintain and repair the demised premises and keep the same in good substantial and efficient repair and working condition and so leave the same at the end or sooner determination of the said term.

Lease agreement, dated 17th January, 1896 between the GER and the GNR for the Ramsey branch.

(d) To well and efficiently manage work stock and use the demised premises and perform and observe all duties and obligations under any and every Act of Parliament for the time being in force relating to or affecting the said demised Railway and the maintenance working management and user thereof and the traffic thereon to which the Great Eastern Company would if this demise were not made be subject or liable in respect of the same and bear and pay all expenses of and in connection with the said Railway and the working maintenance and management thereof and the conveyance of passengers animals and things thereon and including the maintenance of all works for the accommodation protection or enjoyment of lands adjoining the said Railway and the Great Northern Company shall indemnify the Great Eastern Company from the several duties obligations and liabilities aforesaid and from all penalties damages costs and claims in respect thereof AND the said lease shall contain all such other covenants and provisions as are usual in leases of a like nature including a proviso giving to the lessors a right of re-entry in the event of the said rent remaining unpaid for the space of 30 days next after any of the days hereinbefore appointed for the payment thereof or in the event of any breach of the lessee's covenants.

32981

3. THE said lease shall also contain a provision that if any further works shall be required on the said Railway for the accommodation of the traffic thereof or in consequence of any future Parliamentary or Government obligation the Great Northern Company may out of their own funds cause the same to be executed but subject to the previous approval of the Great Eastern Company and that the Great Eastern Company shall at the expiration of the said further term of 21 years pay the Great Northern Company for the works so executed by the Great Northern Company at a fair valuation thereof estimated as their value at the termination of the said lease on the basis of such works being treated as part of a going concern and not as mere land and materials only.

4. THE Great Eastern Company shall be released from their obligation under Clause 3 of the hereinbefore recited Agreement to pay for the further works executed by the Great Northern Company thereunder as aforesaid at the termination of the said recited Agreement of the 15th December 1874 and in lieu of that obligation the Great Eastern Company shall in the indenture of lease to be granted in pursuance of this Agreement covenant that at the expiration of the said further term of 21 years they will pay the Great Northern Company for such further works as shall have been executed by that Company under Clause 3 of the said recited Agreement at a fair valuation thereof estimated at their value at the termination of the said further term of 21 years on the basis of such works being treated as part of a going concern and not as mere land and materials only.

5. THESE presents are made subject to such alterations as Parliament may think fit to make therein but if any material alteration shall be made therein by Parliament either Company may elect to abandon this Agreement.

6. THE Great Eastern Company and the Great Northern Company shall each bear their own respective costs of and incidental to the preparation of this Agreement and of the said lease to be granted in pursuance thereof and of obtaining the sanction of Parliament thereto respectively.

IN WITNESS whereof the Great Eastern Railway Company and the Great Northern Railway Company have caused their respective Common Seals to be hereunto affixed the day and year first above written.

The Common Seal of the Great Northern Railway Company was hereunto affixed in the presence of

F. SHUTTLEWORTH,
Director.

siding to handle the traffic. Authority was given on 3th December, 1897 for the siding to be installed at an estimated cost of £694, to which the GER approved four days later and work was completed the following June.

As the powers for widening the main line obtained in 1891 had lapsed, the GNR sought further powers and authority was duly granted by the Great Northern Railway Act 1898 (61 and 62 Vict. cap. clxv), which received the Royal Assent on 25th July, 1898. Clause 5 included the widening (No. 4) of the main line on both sides between Wood Walton and Yaxley, whilst another clause permitted the crossing of public road No. 4 at Holme station by a level crossing, provided the company erected a footbridge for pedestrians. Clause 40 stipulated provisions for the protection of the junction at Holme with the GER Ramsey branch similar to those set out in the 1891 Act.

In the meantime to the north-east of Ramsey the GER had opened a goods branch from Three Horse Shoes, on the March to Peterborough line, to the village of Benwick, after years of petitioning by local farmers and landowners. Authorized on 30th May, 1895, the line with sidings at roughly half-mile intervals, was opened in two stages from Three Horse Shoes to Burnt House Drove on 1st September, 1897 and thence to Benwick on 2nd August, 1898. Although the promoters thought the new line 'would about bust up the Great Northern Railway' by removing traffic from their Ramsey branch, there is no evidence that the King's Cross Directors were unduly concerned by the new railway. The Benwick branch certainly generated new traffic from the district it served but had little impact on the Ramsey to Holme line, where traffic continued to increase.

Following the satisfactory completion of the cottages for railway staff at Holme, the GNR authorities arranged for another four to be built at an estimated cost of £860, the authority being granted on 5th May, 1899. Early in the new century in 1902 a review of competition against rail for freight traffic revealed there was little threat from road transport but that rivers and canals still conveyed much traffic that could be carried by rail.

Under the revised motive power arrangements it was customary for the early turn New England (Peterborough) men to work the engine of the morning goods train to Ramsey before shunting the wagons and then working all services. The late turn men then travelled passenger to Holme to relieve the early turn men, who returned passenger to Peterborough. On 12th May, 1905 driver H. Southwell of Peterborough worked the early turn and on being relieved at Holme by the late turn men, sent his fireman on ahead to join the down main line train then approaching the down platform, whilst he advised the relieving driver of problems with the locomotive. Meanwhile the down train had stopped at the platform and so to save time, and with complete disregard to regulations Southwell attempted to join the Peterborough train by crossing the up line and jumping on to the offside running board of the coach. As he did so the train started to move. In the confusion Southwell dropped his coat and in trying to retrieve it, he slipped and fell under the moving train. His leg was crushed before the train could be brought to a halt after the train crew heard shouts from the crew of the Ramsey train, who had seen the incident. Unfortunately Southwell died later after the amputation of his leg.

As farming methods improved so traffic on the branch increased. In return for outgoing vegetable traffic, farmers and growers at Ramsey, St Mary's and Holme required an ever-increasing tonnage of manure to fertilize the soil. By the mid-1900s the traffic at Holme had reached such proportions that empty or full wagons were occupying every space of siding and delays were more the rule than the exception. Once again complaints were made to the GNR management regarding delays in transit. The General Manager on the advice of his goods manager deemed it essential to construct additional sidings at the junction station to deal with the interchange of agricultural and manure traffic. Estimated to cost £865, the necessary authority was given on 1st March, 1907, although it was the following year before the sidings were installed and operational.

At about 7.30 pm on 5th November, 1907 during shunting operations at Holme, 43-year-old guard George Gardner was struck by a light engine approaching the station from the Ramsey branch, sustaining two fractured ribs, a scalp wound and lacerations to his left arm as a result of the accident. At the subsequent inquiry the inspecting officer learned that whilst his shunting engine was engaged in another part of the goods yard, Gardner took it upon himself to walk along the track to the up side sidings to couple wagons. After completing the task he walked back towards the station in the six foot spacing between the shunting neck and the Ramsey branch single line but, to avoid a point lever box which stood nine inches above the ballast, he stood foul of the branch line and failed to notice in the foggy conditions the approaching light engine and was struck a glancing blow. The inspecting officer concluded that, as Gardner had only visited Holme sidings on four previous occasions, he was not well acquainted with the track layout and had not taken adequate care as to his safety. However, the guard was absolved from blame as he could not properly ascertain his correct position because of the foggy conditions and more importantly the lack of any illumination to light the yard. The conditions were exacerbated by the falling gradient of the Ramsey single line, which at the point of the accident was nine inches lower than the shunting neck, which caused the ballast in the six foot space between the lines to slope sharply towards the branch. To make matters worse, the inspector found the path in the six foot obstructed in at least six different places by point levers or point rod trunking. For future safety, the inspecting officer recommended that the GNR seriously consider the lifting of the branch line where it ran parallel with the shunting neck, the lowering of the point rods and trunking level with the ballast and the provision of three 'good' lamps in the sidings.

In 1904 a horse-drawn fly was advertised to meet each train at Ramsey station and four years later a horse-drawn coach was provided. If the conveyance was required, application was to be made to Mr F. Miller at the Crown Temperance Hotel. Miller was also the parcels agent for the GNR.

The peaty nature of the subsoil in the area traversed by the Ramsey branch and resultant instability, regularly brought problems to the permanent way. Lineside buildings were also affected, although to a lesser degree but in the spring of 1912 the foundations of the crossing keeper's cottage at Long Drove level crossing were found to be subsiding into the fen and the building was in danger of

St Mary's station and signal box view facing north from the Ramsey Heights to St Mary's road in 1906. The wooden station buildings comprised the booking office and waiting room through the left hand entrance, whilst the station master's accommodation is to the right. An advertisement on the building announces the attractions of Woodhall Spa. The down starting signal protects the level crossing whilst the old carriage body behind the signal box was used as a sack store. St Mary's church complete with spire is to the right. Because of settlement and possible collapse, the spire was later removed.

Author's Collection

collapsing. On 26th July, 1912 authority was given for a new house to be erected on an adjacent site, where the foundations could be placed on firmer ground.

The outbreak of World War I on 4th August, 1914 found the GNR and GER with other British railway companies under Government control. The train services continued to run to pre-war timetables and soon after the commencement of hostilities British farmers and growers were urged to increase the production of grain, vegetables and fruit to offset the deficit of imported foodstuffs caused by enemy action against shipping. Fenland farmers, like their counterparts all over the country, rallied to the call. As a result there was a considerable increase in the number of wagons handled at Holme and the branch stations. Hay traffic was also dispatched as fodder and bedding for horses based at the many military establishments in London, East Anglia and the East Midlands. The Holme to Ramsey line was not as strategically placed as other GNR or GER branch lines and handled few military trains. Military personnel were, however, conveyed by the branch services.

In addition to the root crops, the local farming community was encouraged to produce more grain for the home market. With such an increase in production, Sewell and Son, grain and corn merchants at Ramsey, applied for a siding connection to their new granary, built to the south of the station and adjacent to the up side goods yard, to allow grain to be easily loaded into wagons for dispatch. Sewell agreed to pay the proportion of costs of the siding on his land, estimated at £158, whilst the GNR Engineer estimated the cost on railway land at £113. Provisional authority was given on 7th May, 1915 but as the branch was leased from the GER, the GNR authorities under section 3 of the agreement had to seek the initial permission of the owning company before any expenditure was incurred. The GER Traffic Committee duly sanctioned the provision of the siding on 20th May, and work was completed in time for the harvest.

The resources of the railway companies were severely taxed by the war effort and in December 1916, the Railway Executive issued an ultimatum to the effect that they would only carry on if drastic reductions were made to ordinary services. Locomotive power was desperately short through lack of coal supplies. The Lloyd George Coalition thus agreed to a reduction of passenger train services from 1st January, 1917, but despite this edict the Ramsey branch services remained virtually intact as most ran as mixed trains, with consignments of wagons of produce attached to the passenger carrying vehicles. Local railwaymen answered the call to arms and at least five men left the Ramsey branch during the war period. Some adjustments were made by drafting in relief staff or by the temporary employment of women.

The second term for the lease of the Ramsey line by the GNR was due to terminate on 30th June, 1917 but because of the constraints of war it was deemed unnecessary to seek Parliamentary authority for a future term as the Government had control of all railways. The GNR authorities wrote to the GER and agreed that the lease of the Ramsey branch would continue on the existing terms until such times as the Government control of the railways terminated. The terms ultimately remained in existence until the railway Grouping in 1923.

The festivities of the Armistice were short-lived for, as services attempted to return to normal after the war years, a general railway strike from 28th

Holme station facing north towards Peterborough in 1912. Note the alterations to the station buildings nearest the camera on the down platform. The goods shed is in the background and the island platform with small waiting shelter on the right. Percy East is in the centre of the group on the down platform, whilst Charlie Horinger is on the up side platform.
Courtesy GNR Society

Ramsey North goods yard *circa* 1916, with station and goods yard staff and local merchants posing for the camera. In the background is the original timber goods shed, which was later demolished and replaced by a smaller structure and the 5 ton capacity fixed crane on the loading dock.
Author's Collection

September to 5th October, 1919 paralysed services on the branch, and sowed the seeds for the transfer of traffic away from the railway. The industrial dispute of 1919 heralded the decline of rail borne freight. Farmers and growers realized for the first time that with improving roads, goods could be conveyed by motor lorry using in some cases vehicles purchased second-hand from the army. Short haul journeys were thus made at cheaper rates than charged by the GNR. The door-to-door service was considered more convenient than double handling into and out of railway wagons. The primitive commercial vehicles of the day were not, however, capable of continuous long hauls and the middle and long distance freight traffic remained safely in the hands of the railway company.

In parallel the GNR monopoly of the passenger service was rudely shattered in 1920, when bus services began operating from Ramsey and the surrounding area to Peterborough and Huntingdon. Unlike the GER, which introduced conductor-guard working on the Ramsey High Street branch in 1922, the GNR retained full staffing at stations on the branch from Holme, hoping that falling receipts would prove a short term irritant, and that the offending bus services would quickly disappear from the roads. Difficult times were ahead.

The GNR breakdown crane from New England shed was responsible for attending to any mishap and derailment on the Holme to Ramsey line. Here the breakdown train pulls out of the up sidings on to the main line at Holme in 1922. Nearest the camera is six-wheel guard's brake/tool van No. 272A, whilst the breakdown crane allocated to Peterborough was No. 343A. The bogie van in the centre of the formation contained packing timbers and portable jacks. *Author's Collection*

Clearing out the river wharf on the up side of the line, immediately south of the Ramsey Road level crossing at Holme in the 1920s. *GNR Society Collection*

'J6' class 0-6-0 No. 3589 heads a train of empty wagons on the down main line at Holme in 1925. The 'J6' class was associated with the Ramsey North branch in the LNER and BR eras.
L&GRP

Chapter Four

Grouping to Closure

As required by the 1921 Railways Act, on 1st January, 1923 the GER and GNR were amalgamated with the Great Central, the North Eastern, the North British and several smaller railway companies to form the London & North Eastern Railway. Initially few changes were made by the new owner except that from 1st July, 1923, Ramsey station was renamed Ramsey North and Ramsey High Street station on the Somersham line was renamed Ramsey East. Then on 4th October, the Chief General Manager advised the Traffic Committee that it was proposed to erect a cottage at Whittlesea Mere level crossing, at an estimated cost of £600, for the accommodation of a married platelayer who with his wife, in consideration of free residence and payment of 3s. 0d. per week, would control the gates of the level crossing. This would enable the displacement of the two gatekeepers and result in saving of their annual wages of £234. After allowing for the additional payment to the platelayer and his wife, the net saving was £209.

A seven day rail strike from 20th January, 1924 brought a further decline in railway traffic and passengers who regularly patronized the branch services for short journeys turned to road transport, some never returning to make regular use of the line again. By the summer of 1924 motor omnibus services were operating from Ramsey to St Ives on Mondays and, more seriously for the Ramsey North branch, to Whittlesea on Fridays and Peterborough via St Mary's on Wednesdays, Saturdays and Sundays. These services though infrequent, enjoyed instant success, providing an almost door-to-door service for townsfolk attending local markets for shopping, at a cheaper fare than charged by the LNER.

During the autumn of 1925 the local goods manager complained of the dilapidated condition of the goods shed at Ramsey North. The matter was raised at the meeting of the Works Committee on 26th January, 1926, when the gathering learned the existing timber goods shed dating from 1863 was beyond repair. Replacement by a similar type of shed would cost an estimated £700, whilst a brick structure would cost £800 although then maintenance would be reduced from £16 to £8 per annum. After some deliberation it was agreed to sanction the construction of a brick building and tenders were invited. Two days later, however, the committee had second thoughts and recommended the substitution of a large lock-up shed with awnings on either side at a reduced cost of £559. The estimate included the replacement of the existing goods office, which was considered to be in an inconvenient position, to one adjacent to and forming an extension of the passenger booking office.

Later in 1926 the General Strike brought additional chaos to the railways causing further decline. Officials at King's Cross were far from satisfied with receipts from the branch and whilst goods traffic continued to show satisfactory results, passenger and parcels receipts were poor and fluctuated annually as the results from 1923 to 1928 show:

Holme station, viewed facing north from the footbridge along the double-track main line towards Stilton Fen and Peterborough in the late 1920s. The Ramsey North branch train, formed of 'C12' class 4-4-2T and two ex-GNR coaches has just arrived at the back of the up side island platform. Mail and parcels have been unloaded ready for transfer to the next up main line train but there appear to have been no passengers for destinations south of Holme, although some are waiting for a down train. In the foreground is the simple waiting shelter whilst the Ramsey branch can be seen swinging away across the fens to the right of the picture. The up side reception sidings are beyond the curve of the branch, running parallel with the up main line. *Author's Collection*

Ramsey North station in the late 1920s. 'C12' class 4-4-2T No. 4516 is running round the branch train which is formed of ex-Great Northern Railway six-wheel stock. The 'temporary' station sufficed until the closure of the line and was quite open to the public at all times. On the left is the timber goods shed, which because of deteriorating condition, was soon to be replaced by a smaller structure. *Author's Collection*

	Passengers	Passenger receipts £	Parcels receipts £	Season ticket receipts £	Total £
1923					
St Mary's	13,759	561	52	51	664
Ramsey North	24,903	2,138	383	412	2,933
Total	*38,662*	*2,699*	*435*	*463*	*3,597*
1924					
St Mary's	13,844	611	63	–	674
Ramsey North	26,887	2,358	311	141	2,810
Total	*40,731*	*2,969*	*374*	*141*	*3,484*
1925					
St Mary's	14,288	542	60	2	604
Ramsey North	28,834	2,496	353	68	2,917
Total	*43,122*	*3,038*	*413*	*70*	*3,521*
1926					
St Mary's	12,140	394	58	4	456
Ramsey North	22,868	1,994	314	86	2,394
Total	*35,008*	*2,388*	*372*	*90*	*2,850*
1927					
St Mary's	15,211	391	54	3	448
Ramsey North	25,585	1,916	399	109	2,424
Total	*40,796*	*2,307*	*453*	*112*	*2,872*
1928					
St Mary's	11,033	358	54	5	417
Ramsey North	23,620	1,951	667	48	2,666
Total	*34,653*	*2,309*	*721*	*53*	*3,083*

In the six years the average number of passengers booking weekly at the branch stations fell from 743 to 666, reflecting the improvement of the local bus services and the decline of the travelling public's reliance on the Ramsey North branch. The LNER Directors had for long been concerned with the poor returns earned from passenger traffic where competitive bus services running on roads paralleling the railway had seen further improvements. At a special Board meeting it was resolved to stem the loss of receipts by introducing 20 railway company-operated bus services with the purchase of 60 motor buses. One of the centres chosen for the new services was St Ives, where it was envisaged routes would operate to Holme via Ramsey and Huntingdon via Godmanchester, using two 20-seat vehicles and to Ely via Sutton, using four 32-seat vehicles, in competition with other services. Allocations would, however, be subject to amendment in the event of the LNER coming to terms with local bus proprietors. The Engineer estimated the cost of providing garage accommodation at St Ives at £5,000 exclusive of lighting and petrol pumps. On 29th November, 1928 authority was given for the outlay on the garage at St Ives but no further developments were made for the company was negotiating financial and operating interests in the existing local bus companies. Thus by the late 1920s the LNER suffering a

Holme station, viewed from the down platform facing south in the late 1920s showing Harper's suspension footbridge and the Ramsey branch 'C12' class 4-4-2T taking water from the column beyond the level crossing. Note the schoolboys 'trainspotting' on the footbridge and the period Bovril advertisement to the right, one of several published at this period with a railway theme.
Stations UK

The epitome of the branch train. 'C12' class 4-4-2T No. 4516 stands at Ramsey North in 1929 with her train composed of ex-GNR four-compartment six-wheel brake/third, five-compartment six-wheel composite and five-compartment, six-wheel third. The GNR class 'C2' later LNER class 'C12s' were regularly employed on the branch after displacement from the London suburban services until 1931.
The late J.E. Kite

massive traffic recession and the sustained losses in revenue deemed it necessary for the management at Marylebone and Liverpool Street to seek economies. Many branch lines in East Anglia were closely investigated as to their viability in the future passenger railway network. The dwindling receipts of both Ramsey branches made depressing reading, and the luxury of two stations in a town producing an average of 834 passengers weekly, made reductions inevitable.

After weighing up the advantages and disadvantages it was decided the Ramsey North route with 667 passengers booking weekly, compared to 167 passengers from the East, proved the more viable of the two branches, with three times the receipts of the Ramsey East line. On 3rd September, 1930 the Divisional General Manager, Southern Area advised the LNER Board that as the Ramsey East branch was working at a loss it was proposed to withdraw the passengers services from that line on and from 22nd September, 1930. The Board readily agreed when it was realized that the withdrawal of the service would save an estimated £2,669 annual working expenses. After the loss of receipts, estimated at £540, the net saving was £2,129. The Eastern National Omnibus Co., an associate of the LNER, provided an alternative service to Warboys and St Ives, and the last train ran on Saturday 20th September, 1930. Freight and parcels continued to be handled at Ramsey East and Warboys, whilst the occasional excursion train continued to cater for day trippers wishing a few hours by the sea at Clacton or Hunstanton.

Any complacency regarding the future of the Ramsey North branch passenger services was rudely shattered on 22nd December, 1930, when the Divisional General Manager, Southern Area reported to the LNER Board that, although the line carried an appreciable amount of goods traffic, the gross value of passenger traffic local to or originating or terminating on the Holme to Ramsey North branch in 1928 only totalled £5,528, with no indication of any increase. It was not considered desirable to withdraw the passenger train service completely as the first service trip over the branch was a goods train, and after arrival at Ramsey the locomotive worked passenger services up to the 10.15 am ex-Holme. In the interests of economy it was proposed on and from 2nd February, 1931, to retain the working and then cancel all passenger services on the line after the 10.15 am down train. The Board sanctioned withdrawal on learning that the estimated annual savings of £2,238, offset by the annual loss of passenger train earnings of £1,179, would realize a net saving of £1,059 per annum. The necessary notices were duly posted at the branch stations and neighbouring stations on the main line, whilst arrangements were made with the Peterborough Electric Traction Co. to provide an alternative bus service between Ramsey and Holme. The LNER agreed to contribute towards the cost of the replacement road service. Compared with the railway closures of the 1950s and 1960s little opposition was made by local inhabitants. The Peterborough Electric Traction Co., which initially operated the replacement service, was soon taken over by the Eastern Counties Omnibus Co. (ECOC). Based at Norwich and serving a vast area incorporating the counties of Cambridge, Norfolk, Suffolk, Huntingdon and Northampton, the company was another subsidiary of the LNER. Tilling 'B10A' type buses operated the initial replacement services to Holme and connections were maintained with main line rail services.

As well as investigating the passenger revenue, the LNER authorities also delved into the situation regarding freight traffic, and in particular the standing of the rival forms of transport. The virtual elimination of waterborne competition had been replaced by the more serious threat of the road vehicle. The report on the waterways in the Ramsey area stated:

> In the vicinity of Holme, the Middle Level Drainage Commissioners do not consider navigation a primary consideration and the water level is so low that during the past two or three years traffic has not been boated to Holme Wharf in appreciable quantities, though we get it both by road and by boat at St Mary's. At Ramsey North very little traffic is boated for either the local mills or rail transit ... [it concluded] ... bearing in mind the fact that roads in country districts are now in much better condition, whilst many farmers possess motor lorries of their own and there are many road haulage companies in existence, it seems to us that the fear of water competition is nothing like the same as in 1902, our principal enemy now being road transport.

By the late 1920s sugar beet cultivation was established in the fenlands and during the winter months a considerable tonnage was loaded at St Mary's and Ramsey North for conveyance to the beet factory at Peterborough. Beet traffic tonnages increased during the 1930s and made a welcome addition to the branch freight receipts, which were in danger of reduction as other commodities were lost to road transport, notably cattle and horse traffic. With a view to effecting economies and reducing operating costs, authority was given on 9th January, 1936 to abolish the signal boxes at St Mary's and Ramsey North, together with associated signalling. The two signal boxes, which were manned by porter/signalmen, were in need of extensive repairs. The points leading from the main line to the sidings could be worked from a ground frame at each place, with levers released by Annett's key on the Train Staff. The conversion work was estimated to cost £223, offset by £10 recovered materials and £82 for the net cost of renewals required immediately, which were not now required, leaving a net cost of £131. The estimated original cost of the displaced works was £752. The scheme was adopted and permitted a reduction in maintenance and renewal charges of £91 per annum. The work was completed later in the same year when the signal boxes were duly abolished on 2nd June.

The main line bridge reconstruction programme authorized in 1935, amongst other things proposed the reconstruction of public footbridge No. 170 at Holme station using two serviceable girders at a cost of £550. However, when work commenced in the spring of 1936 it was found the girders caused problems with signal sighting. It was therefore necessary to order two new girders of suitable design to obviate the problem, which increased the cost to £950. The work was authorized on 25th June, 1936, with the contract for the supply of steelwork being awarded to Wright, Anderson & Co. who tendered at £463 7s. 9d.

By 1936 the Eastern Counties Omnibus Co. was operating daily services from Ramsey to Peterborough and Whittlesea on Fridays only, St Ives on Mondays only, Huntingdon on Saturdays and Sundays, March on Wednesdays and Saturdays, whilst the Ramsey East branch replacement bus service to Somersham ran daily except Sundays. Blue Bird buses were also operating in direct competition against the LNER from Holme to Peterborough on

Holme station and level crossing facing south in 1936 with a down main line goods train approaching. The Ramsey North branch trip engine is shunting in the down south end sidings beyond the signal box. To the left is the small timber waiting shelter on the up side island platform. Beyond the level crossing is the water column and storage tank. In the far background two LMS wagons stand in the wharf siding. Notice the absence of the former footbridge, which was later reinstated. *Author's Collection*

Wednesdays, Saturdays and Sundays. As a result, less reliance was placed on the early morning passenger train service on the branch and by 1939 the rail replacement bus service to Holme was operating as one round trip from Ramsey to the junction and back in the afternoon. Other bus routes continued as before, the largest drain on the branch traffic being the daily Ramsey to Peterborough services.

As war clouds gathered over Europe materials were consigned to Ramsey for onward transit by road to local airfields at Wyton and Upwood. Upwood airfield to the south-west of Ramsey was originally established in 1917 as a Royal Flying Corps base for two night training squadrons but with the ending of World War I the squadrons were disbanded and the airfield reverted to farmland. The base was rebuilt in January 1937 with the threat of hostilities and became the home for two light bomber squadrons.

On the outbreak of World War II the LNER merged with other major railway companies under the Railway Executive Committee. To safeguard against air raids especially at night, station and signal box lamps remained dimmed and staff utilized shielded handlamps to attend to train and shunting duties. As a precaution against enemy infiltration, station name boards were also removed and stored in lamp rooms and other inconspicuous places. Once hostilities commenced additional passenger trains occasionally ran across the branch conveying military personnel. Most personnel movements were, however, effected via Peterborough and thence by bus. The agricultural nature of the freight handled on the branch became of the utmost importance as vital provisions of home grown food, grain and vegetables were dispatched to London and the provincial markets. Rationing of petrol brought the withdrawal of many motor vehicles from the local roads leaving the railway to convey the former road borne produce. During this period ammunition wagons were sent to Ramsey North conveying shells and other items for local airfields, the ammunition being transferred to road vehicles for onward conveyance under the cover of darkness. The ammunition was usually conveyed in open wagons, sheeted over to conceal their deadly cargo, although the prominent red flashed labels advised 'shunt with great care' and 'place as far as possible from the engine, brake van and wagons labelled inflammable'. Some consignments of aviation fuel were also conveyed to Ramsey North. In late 1943 concrete runways were laid at Upwood with much of the materials being conveyed by rail to Ramsey North then on to site by road. No. 2 Pathfinder squadron with 'Mosquitos' and 'Lancasters' of Bomber Command were resident and during the operational period from February 1944 to the end of the war flew 6,000 sorties, including 36 consecutive night attacks on Berlin by the 'Lancaster' bombers. The Luftwaffe, however, retaliated on a number of occasions and in 1940 one person was killed. At Wyton, Bomber Command flew 'Halifax' aircraft on bombing raids.

After the war effort the railways were totally run down. Questions were raised in Parliament regarding the poor state of the transport system when many trains were cancelled because of defective rolling stock and equipment. The Ramsey North branch was not immune and on several occasions the goods and passenger services were cancelled through lack of motive power. For the benefit of Royal Air Force personnel travelling on a weekend Furlough during 1945 and 1946, a passenger train ran from Ramsey North to Peterborough on Friday evenings and returned from Peterborough on Sunday evenings.

From 1947 the tide was again changing against the LNER as parcels and freight traffic gradually transferred away to road haulage. With the easing of petrol rationing additional buses returned to the roads and farmers and growers took advantage to convey their produce from door to door by lorry instead of double handling into and out of railway wagons. The railway officials at Liverpool Street, King's Cross and Marylebone investigated the viability of each line in their area and noted the poor receipts earned by the Ramsey North passenger services. A survey in 1946 established that in 1931, 14,000 passengers booked through Ramsey North station whereas in 1946 the figure had dropped to 2,665, an 80 per cent reduction. The arrangement for morning-only rail services and one return bus journey in the afternoon attracted few customers, with the highest loaded train conveying a mere nine passengers, and it was decided the loss making arrangements could not continue. On 3rd July, 1947 the Divisional General Manager advised the Traffic Committee of the proposal to withdraw the passenger train service between Holme and Ramsey North, and the termination of the working arrangement with the ECOC. The annual payment made by the LNER to the ECOC towards the cost of providing an alternative bus service based on the 1946 figure was £250 per annum, but the withdrawal of all passenger facilities from the branch was expected to produce a net annual saving of £2,700. Needless to say the committee readily agreed to the withdrawal of services, as latterly the main source of revenue came from servicemen returning late to their airfields after weekend leave. Notices were duly posted at stations in the neighbourhood and in local newspapers that the passenger service would be withdrawn on and from 6th October, 1947, with the actual last train running on the previous Saturday.

Unlike the closures of the 1930s, when there was mute response, Ramsey inhabitants voiced their opposition to the withdrawal of services. Complaints were made that the town would become isolated with people 'unable to get out of the place'. The vice-chairman of Ramsey Urban District Council blamed the LNER for pruning off uneconomic parts of the system without regard for their obligations to the travelling public. The lack of convenient connections at Holme and the abysmal service provided by the few remaining passenger trains, and the one afternoon bus in each direction, hardly encouraged people to use the service. Advocating subsidy, he concluded 'we contend the company are supposed to give a service to the public and they should regard the receipts only from the point of view of the final result over the whole system'. Ramsey businessmen met on 24th September, 1947 in a last ditch attempt to save the service and David Renton, the local MP, launched a campaign to keep the line open. The businessmen wrote formally to the LNER management stating that if a reprieve were made the line would be better supported, particularly in view of the petrol rationing.

The arguments for retention of the passenger services fell on deaf ears at King's Cross, and at 10.15 am on 4th October, 1947 the last passenger train departed from Ramsey with three passengers, all described as railway enthusiasts, two of whom had travelled from London. Ramsey people were conspicuous by their absence and even the booking office was closed and shuttered. On arrival at St Mary's the trio were permitted to purchase tickets from the booking office, which was opened especially for the occasion, to commemorate their journey. The *Huntingdonshire Post* truly reported, 'Little interest was shown in the event'. The day after closure, a letter from the LNER was read out at a meeting of the Ramsey UDC, confirming that the passenger services on the line had been losing money for a considerable time and as there was no prospect of it paying its way, trains had been withdrawn to save an estimated £2,450 and 200 tons of coal per annum. Another source of feeder traffic lost was the horse-drawn barges delivering vegetable traffic to Holme Wharf where the produce was transferred into rail wagons. The barges then returned with coal for the outlying communities. The silting up of the waterway forced abandonment in 1947.

From 1st January, 1948 the railways were nationalized and the former Ramsey Railway came under its fourth ownership, that of British Railways (BR). Few alterations were made and the branch retained its GNR/LNER atmosphere until closure. Locomotives working the services soon lost their 'NE' or 'LNER' identity from the tender sides, to be replaced by the legend 'BRITISH RAILWAYS'. Wagons and goods brake vans were progressively painted in the new corporate BR livery. The new owner, however, found the task of attracting traffic to the branch increasingly difficult as road conveyance of freight went from strength to strength.

On 16th April, 1952, 65-year-old J.T. Short, who was employed as a lengthman on the Ramsey North branch, sustained injuries in an accident at Holme station. He was working with the local permanent way gang between the platforms repairing the up main line track, and at 2.55 pm had just lowered a jack, which was being used in the six foot space, when the handle was struck by the axlebox of a wagon on a freight train passing on the down line. He sustained injuries to his knee and was incapacitated for one month. An inquiry was ordered by the BoT and was conducted by their inspecting officer J.A. Sinclair.

During his investigation, Sinclair visited Holme and was surprised to find that the space between the up and down main line tracks never exceeded 5 ft 3 in., whilst at the country or north end where the gang was at work, the spacing was reduced to 5 ft 2 in. As the handle of the type of jack in use extended to a length of 4 ft 2 in. from the rail, when horizontal, and about three minutes were required to pack two sleepers, a sharp watch had to be kept for trains approaching on the adjacent track. At Holme, because of the restricted space between the tracks, it was usual for the jack to be used in the four-foot way as far as practicable. In this case, however, Sinclair was told a connection at the north end of the station rendered it necessary for the implement to be used in the six-foot space.

At the inquiry Short advised the inspecting officer that he, with two others, had been engaged repairing the up main line track for much of the day and they were joined by ganger Albert Brudenell after he had completed the daily examination of the Ramsey branch. Short was operating the jack, a duty he had frequently undertaken when working with the gang, and in accordance with

Holme station viewed from the signal box facing north in the early 1950s. The Ramsey branch goods train hauled by a 'J6' class 0-6-0, and formed of a covered van and brake van is standing at the back of the island platform, by now devoid of a waiting shelter, as an 'A4' class 4-6-2 speeds through with an up express. *Author's Collection*

Ramsey North station from the buffer stops in April 1957, facing west across the flat fenland. Note the difference in height where the platform was lengthened. The wagons behind the platform stand on the cart road whilst from the platform road the lines to the right are the run-round loop, middle road and shed road. Ramsey North was always considered to be a bleak station especially when the wind was blowing from the north or the east, with little shelter provided for waiting passengers.   *The late H.C. Casserley*

the normal procedure kept watch for approaching trains, which could be seen for a considerable distance in each direction. Several lifts had already been made when freight trains were seen to be approaching on both up and down lines. As the train on the up line was some three-quarters of a mile away it was decided there was time for another lift, although it was realized this would probably involve lowering the track as the down train passed. When the packing of the sleepers was completed, Short lowered the jack, and stood clear in the six foot on the north or Peterborough side of the handle. As the train passed the handle was struck by the vehicle and cannoned into his knee causing the injury. Sinclair then questioned other members of the staff who concurred with Short's account of the incident.

In his summing up, Sinclair considered that in view of the local conditions Brudenall, who was in charge of the work, took an unnecessary risk allowing the track to be raised, when he should have been aware that it would have to be lowered again while the down train was passing. The inspector was sympathetic to the situation as both tracks through Holme station were seldom unoccupied for any length of time during working hours and there was a great temptation to take every reasonable opportunity to lift the track between trains. The question of using an alternative type of jack was considered but found to be impracticable at Holme. Sinclair concluded, 'It would appear therefore that until such times as Holme station is reconstructed more in keeping with modern requirements, the present method of working will have to continue.'

By the mid-1950s dwindling traffic to and from the town of Ramsey was causing concern at Eastern Region headquarters. Little traffic was handled at the East station goods yard, and indeed on many occasions the daily freight train only ran as far as Warboys before returning to Somersham. Deteriorating permanent way on the North branch necessitated expensive renewals and the

'C12' class 4-4-2T No. 67380 stands at Ramsey North station with the Railway Enthusiasts' Club 'Charnwood Forester' railtour train on 14th April, 1957. *R.M. Casserley*

'C12' class 4-4-2T No. 67380 standing at Ramsey North with the 'Charnwood Forester' railtour train on 14th April, 1957. *A.E. Bennett*

need for economy called for a closer inspection of receipts. After investigation it was decided all Ramsey traffic could be routed via the North station and the Eastern Region subsequently announced the closure of Ramsey East as a public goods siding on and from 17th September, 1956. Ramsey North goods yard, however, only received a slight increase in sugar beet and coal traffic as a result of the closure. Despite the increase in vegetable traffic being transferred to road transport for conveyance to Covent Garden and Spitalfields markets in London, lengthy loads were still being handled on the Ramsey North branch. In season the last up branch train often departed from St Mary's with a tail load of up to 50 wagons. Further rationalization made in the mid-1950s required trainmen to open and close the gates over the public roads at Whittlesea Mere, St Mary's Fen and St Mary's station level crossings. Warning signs were placed 440 yards on each side of the crossings to advise the staff accordingly. St Mary's was also reduced to the status of an unstaffed siding from 2nd May, 1960.

The Ramsey North branch received the attention of railway enthusiasts on 14th April, 1957, when the Railway Enthusiasts' Club 'Charnwood Forester' railtour visited the line. Originating at King's Cross, the train was hauled by an 'A4' class 4-6-2 tender locomotive for the journey to Holme, where New England-based 'C12' class 4-4-2T No. 67380 took over the five-coach train for the run across the branch. The train paused at St Mary's so that photographs could be taken before continuing to Ramsey North. After running round, the 'C12' locomotive and its train then returned to Holme to enable the special to continue its tour, which included Stamford, Market Harborough, Glenfield, Shepshed, Nuneaton, Bedford and Hitchin. Another 'C12' class engine, No. 67357 hauled the train from Essendine to Stamford

The many closures of branch lines and stations, which began in the early 1950s, continued as ongoing investigations by British Railways highlighted more and more uneconomic candidates for closure. With the loss of the Ramsey North passenger traffic, Holme station had over the succeeding years seen a decreasing number of passengers, and except for the occasional tripper to London and the few regular commuters to Peterborough, days went by without a ticket being issued. The case for retention was not helped by the meagre service offered by the main line trains to and from King's Cross, which totalled three on Saturdays excepted and two on Saturdays only, whilst no trains called on Sundays. Investigation proved the station a financial liability and passenger services were subsequently withdrawn on and from 6th April, 1959.

The infamous Beeching Report *The Reshaping of British Railways* was published in 1963 and although specifically referring to passenger traffic, maps were included within the document showing density of freight traffic and distribution of freight tonnage. The figures registered for the Ramsey North branch were far from encouraging, for the line carried under 5,000 tons per week, whilst St Mary's dealt with under 5,000 tons per annum and Ramsey North under 25,000 tons per annum

Steam traction gave way to diesel on the branch freight services from early 1964 but before its passing, a New England-based London Midland Region class '4MT' 2-6-0 hauled the first of two special trains across the branch organized by the Cambridge University Railway Club in March of that year. For

LM class '4MT' 2-6-0 No. 43084 shunting vans in the middle road at Ramsey North in 1961. In the background are the loading dock and 5 ton capacity fixed crane. To the right the shunter is operating the east end ground frame. LM class '4MTs' from New England shed worked the branch goods services in the last years of steam traction, although the local permanent way inspector frowned upon the use of these locomotives as he thought they were too heavy for the timber underbridges on the branch.
The late Dr Ian C. Allen

Holme station facing north showing the signal box and footbridge No. 170 spanning the main lines and Ramsey branch in the late 1950s. On the down platform is the station master's house, slightly altered from that shown in the 1875 view. The goods shed is beyond the station buildings. The Ramsey North branch brake van stands at the back of the up island platform whilst the independent Ramsey branch crossing gates are closed across the line. The up Ramsey branch GNR somersault signal can be seen below the steps of the footbridge protecting the crossing. *Author's Collection*

General view of Ramsey North station and goods yard, facing towards Holme in 1961. To the left is the granary and associated up side sidings. The station buildings, always considered to be 'temporary' but never replaced, are in the centre and the goods shed to the right. Note the ornate gas lamp in the foreground. *Author's Collection*

The road approach to the wooden station building at Ramsey North, which at one time housed the booking office, waiting rooms, porters' room, station master's office and toilets. Townsfolk were complaining of the 'temporary' structure and asking for their replacement as early as 1879. The GNR declined to provide a better station as passenger receipts from the line 'barely warranted the expense'. The original structure, with slight alterations, remained until the closure of the branch. *Author's Collection*

Members of Cambridge University Railway Club made a brake van trip across the Ramsey North branch just before the withdrawal of services. Here the 350 hp diesel-electric shunter No. D3449, adorned with a headboard, waits at the platform at Ramsey for members to rejoin the train. *Author's Collection*

On 2nd May, 1960 St Mary's was reduced in status to an unstaffed goods station. The station buildings were then demolished and track rationalized. This 1964 view shows the main single line and level crossing. The siding to the goods dock has been removed and only the brick-built transit shed and back road are extant. *Author's Collection*

Ramsey North station facing west in the period just before withdrawal of services. Over the years considerable rationalization of infrastructure had taken place as traffic declined. Only the platform road and truncated run-round loop remain with hand-operated points. The goods shed and fixed crane have gone and the station site bears signs of desolation. A 350 hp diesel-electric shunting locomotive waits at the platform with the branch goods ready for the return run to Holme. *Author's Collection*

The east end of Ramsey North station showing the timber building just before closure of the line. Diesel-electric shunter No. D3449 waits at the platform to depart with the branch goods to Holme. By this date traffic was sparse, hence the single covered wagon and goods brake are the consist of the train. In the left background is granary road, whilst to the right is the former wagon weighbridge office, later used as an office by the station master and then the commercial agent. *Author's Collection*

The remains of St Mary's station with brick goods transit shed, view facing Holme in 1983.
Author

the trip two additional brakevans were added to the formation of the Peterborough to Ramsey North goods to convey the enthusiasts. The second trip, a week later, was worked by a 350 hp diesel-electric shunting locomotive, No. D3449.

Dwindling receipts and the loss of sugar beet and brick traffic brought the closure of the Warboys to Somersham section of the Ramsey East branch on and from 13th July, 1964, the same day as the complete closure of the neighbouring Three Horse Shoes to Benwick goods line. On the Ramsey North branch goods traffic had settled down to maintain an even flow with imports of coal (4,000 tons in 1966), coke, seed potatoes, fertilizers and the occasional cattle. Exports consisted of vegetables including potatoes of which 20,000 tons were dispatched in 1966, sugar beet and fruit. In the autumn of 1966 the BR authorities carried out rationalization in connection with a repainting programme for St Mary's and Ramsey North stations. After considering the requirements of traffic, it was decided to demolish the goods shed at St Mary's but repaint the station buildings, then used as living accommodation by the foreman-in-charge. In fact the former was retained and the latter subsequently demolished. At Ramsey North the old sack store had been demolished and it was decided to retain the goods shed and station buildings, which were eventually repainted at a cost of £774. The yard inspector's office was demolished with the occupant transferring to the station buildings. Considerable rationalization was also achieved at Holme where the old timber hut in the south yard, known as Bibby's store, the former weighbridge office used by the signal and telegraph department, the north yard stables used by the district engineer, the goods shed and two old wagon bodies were all demolished. The occupants were transferred to the former down side station buildings, which were also used as the goods office. This structure, together with the former porters' room and toilet, were repainted at a cost of £289 the following year.

For almost a further decade the branch continued to serve the locality despite the ever-increasing use of motor transport. By the early 1970s the tide had turned completely against the railway and the ever-dwindling traffic receipts concerned the Eastern Region authorities at King's Cross and York. Freight facilities were withdrawn from Holme on 31st October, 1970 and deteriorating permanent way and the need for expensive renewals, together with the gradual subsidence of the trackbed in many places, caused management to seriously question the future of the branch. The investigations only confirmed the inevitable and in April 1972, the Eastern Region announced the closure of St Mary's goods yard with effect from 28th February, 1972 and just over a year later, withdrawal of freight from Ramsey North on and from 2nd July, 1973, with alternative facilities available at Peterborough. In the end traffic was exceptionally light, although the first morning down trip continued to carry a small amount of fish traffic for Ramsey shops until the last day of working.

Within months of closure, on 11th February, 1974 the East Anglian Economic Planning Council, reviewing the future economic strategy for East Anglia, recommended the possible reduction of freight by road transport, by calling for a careful examination of the need for rail freight routes to remove traffic from road back to rail. Included in the proposal was the Holme to Ramsey North line but early in the investigation the branch was eliminated from the plan, and with it all hope of resurrecting the erstwhile Ramsey Railway.

The remaining buildings on the branch were soon vandalized and after some months of inactivity contracts were placed for the removal of the track and fixed assets. Rails were cut up into convenient lengths and loaded on to lorries, which gained access to the former railway trackbed by using the numerous level crossings. Work commenced early in September 1974 at Ramsey, where the site was razed to the ground, and continued slowly towards Holme. Items of pointwork at Ramsey and St Mary's, together with two of the ex-GNR somersault signals on the branch at Holme, were purchased by the Peterborough Railway Society for use on the preserved Nene Valley Railway. The signals, in modified form, still protect the level crossing at Wansford station. The other somersault signal was re-erected in the garden of a house near Holme level crossing.

Returning to the replacement bus services, in the 1940s and 1950s the Eastern National and later Eastern Counties standardized their fleets and Bristol 'Ls' and later 'SCs' and 'MW' single-deck buses, with Eastern Coach Works bodies, operated the routes from Ramsey. The company later operated Bristol 'LH/RELL' and Leyland 'National' type single-deck vehicles but the service operated by Eastern Counties no longer connected Ramsey with Holme. In the late 1980s one bus ran each way on Wednesdays only on route 133 connecting Bluntisham with Peterborough via Ramsey and St Mary's. Ramsey, St Mary's and Peterborough were also connected by Eastern Counties route 230 operating about five journeys each weekday with additional journeys on Saturdays.

The trackbed of the Ramsey North branch can vaguely be traced although most sections have been acquired by local farmers and ploughed in. The ubiquitous railway fence is in some cases the only evidence of the former route. Initially Inter-City 125 and later class '91s' and now London and North Eastern Railway Azuma Inter-City express units speed past the site of Holme station, where the former junction is just discernible by a few earthworks on the up side where the branch curved away to the east. Remains of the old wharf lie south of the level crossing. On the down side the loading dock and station cottages are evidence of former GNR ownership. A new signal box controlling the level crossing but under the supervision of the Peterborough power signal box replaced the former GNR Holme signal box, which was demolished on 2nd June, 1976. At the same time barriers replaced the gates. The former crossing cottages are extant where the road bisects the former branch trackbed at Whittlesea Mere and Long Drove. At Ramsey St Mary's the old brick goods transit shed was demolished and a bungalow built on the site of the station. A hump in the Ramsey St Mary's to Ramsey Heights road provides the only evidence of the former level crossing. Raised earthworks of the trackbed herald the approach to Ramsey North where the former station and goods yard site had been levelled and used for lorry and material storage by a local firm before redevelopment. The Railway Inn across the road, however, continues to offer refreshments to travellers.

Chapter Five

The Route Described

Holme station, 69 miles 29¼ chains from King's Cross on the ex-GNR main line to the north, was the junction for the Ramsey North branch. Situated immediately north of Long Drove Road level crossing No. 87, the station had three platforms, the down side, 360 ft in length, serving the down main line being host to the typical GNR-style country station buildings including booking office, waiting room, gentlemen's and ladies' toilets. The up side platform was an island, the west face 305 ft in length serving the up main line whilst the east face, 280 ft in length, was normally used by the Ramsey branch trains. The island platform was provided with a small waiting shelter for passengers and access between the up and down side platforms was by a foot crossing across the tracks. On the London side of the level crossing a footbridge was provided to enable pedestrians to cross the line when the level crossing gates were closed across the road for the passage of a train. The original suspension footbridge provided in 1879 was removed when plans were prepared to widen the main line to four tracks in the late 1920s. However, when widening was abandoned a second bridge was erected in 1936, and again removed in 1958.

Goods facilities at Holme included four sidings and a run-round loop on the down side of the main lines, south of the level crossing. These sidings included the 1,000 feet-long reception road, known locally as Temporary road which later in World War II became part of the down goods line from Connington North. The other sidings Nos. 1, 2 and 3 were 300 ft, 250 ft and 200 ft in length respectively. On the up side of the main lines the Ramsey branch continued past the island platform in a southerly direction over the combined level crossing, later converted to an independent level crossing, to serve Holme Wharf. Here the siding was 470 ft in length, with a short spur leading from a wagon turntable. These were provided to facilitate the interchange of traffic from horse-drawn fen lighters and barges using New Dyke, which was navigable almost to Holme station.

North of the station on the down side of the line was the goods yard for domestic traffic, served by the 250 ft coal road and 300 ft shed road serving the goods shed together with the 700 ft-long shunting neck road which ran alongside the down main line. On the up side, running parallel to the main line north of the station and located north of the curve of the Ramsey branch were six sidings, three reception lines 2,554 ft, 871 ft and 844 ft in length respectively. The other three sidings, 316 ft, 238 ft and 283 ft in length were used for carriage and wagon storage and interchange traffic to and from the branch.

Points and signals for the main line and the Ramsey North branch were controlled from Holme signal box located on the down side of the main line immediately south of Long Drove Road level crossing. The level crossing for the main line could be worked independently of the single line Ramsey branch crossing, and all were originally worked by a gatekeeper, but later a wheel operated by the signalman opened and closed the gates.

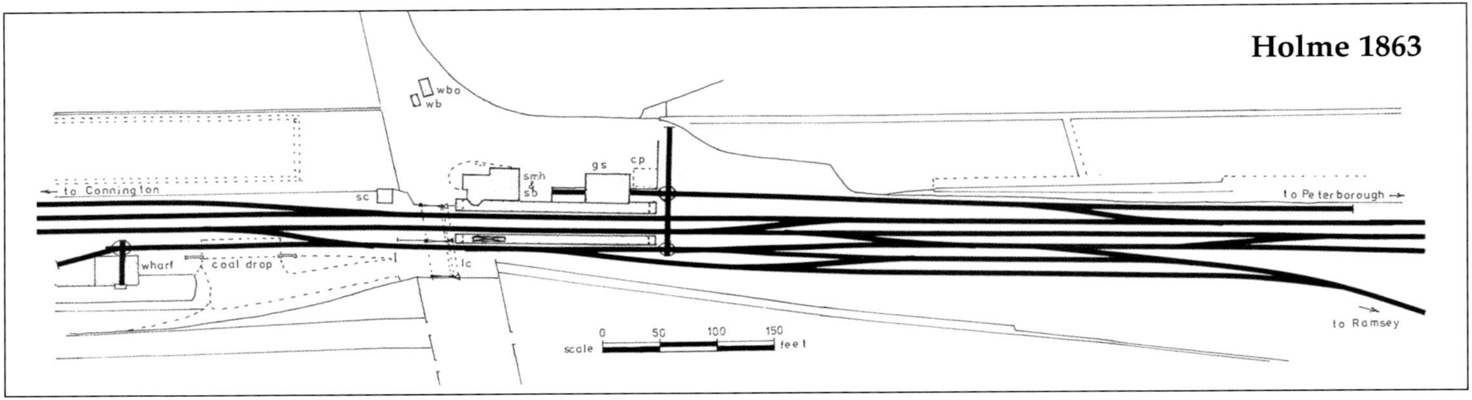

Holme 1863

to Peterborough
to Ramsey
to Connington

Key to station plans

cb	Coach body
ckh	Crossing keeper's house
co	Coal office
cp	Cattle pens
es	Engine shed
fb	Footbridge
fc	Fixed crane
gf	Ground frame
go	Goods office
gs	Good shed
lc	Level crossing
ld	Loading dock
lg	Loading gauge
lr	Lamp room
ls	Loading stage
oc	Occupational crossing
pwh	Permanent way hut
sb	Station building
sc	Signal cabin
smh	Station master's house
sp	Signal post
wb	Weigh bridge
wbo	Weigh bridge office
wc	Water column
wt	Water tank

View of the mixture of brick and timber buildings on the down platform at Holme in 1957, with the up island platform in the foreground. The Ramsey North branch track is nearest the camera with the points leading to the run-round loop.

Author's Collection

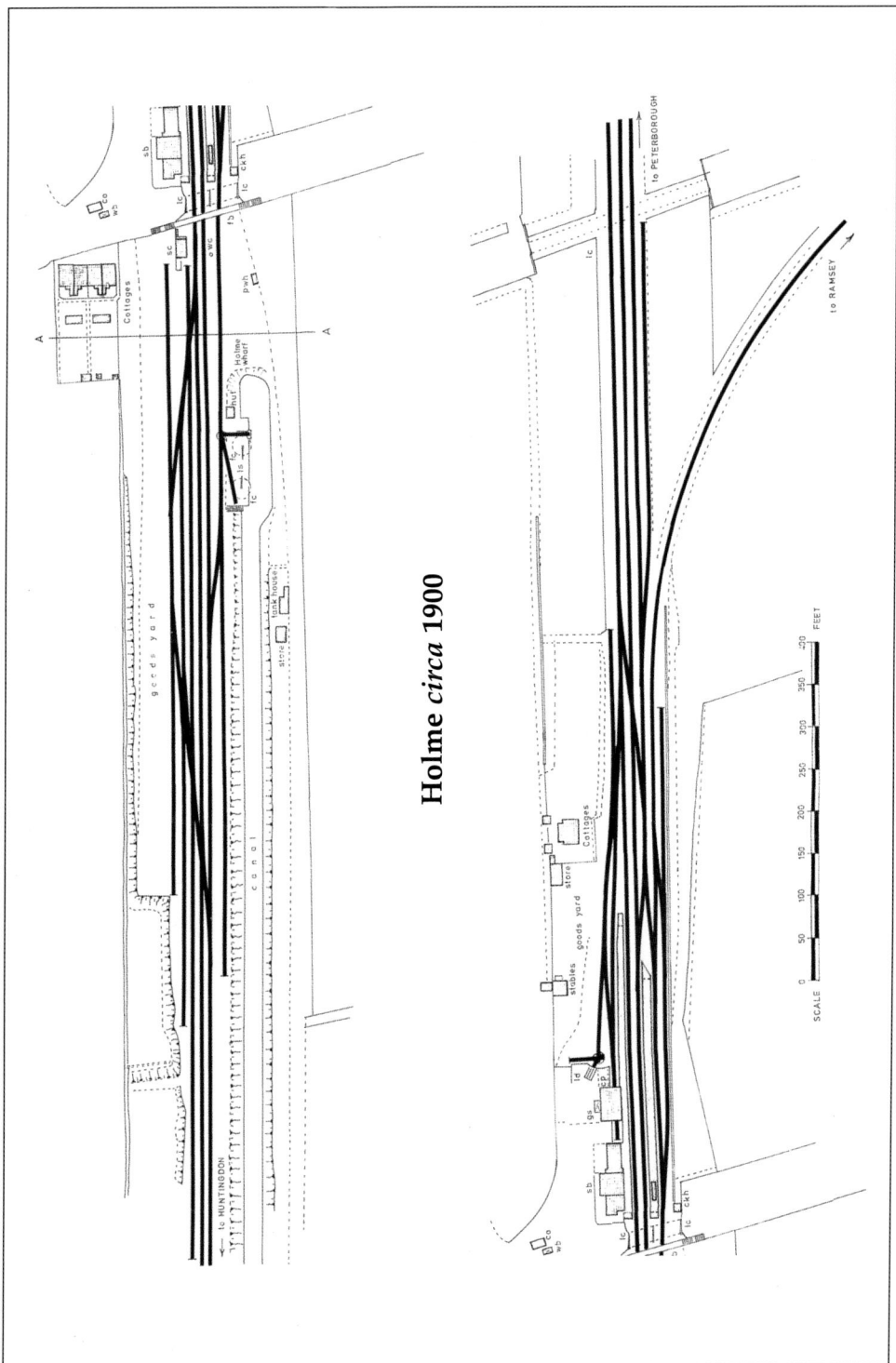

Holme *circa* 1900

Holme station from the south end of the up side island platform in 1955. Beyond the station building on the down platform is the goods shed. Note the oil lamp cradle to the right and the absence of the shelter on the up platform. *Stations UK*

Holme station and main line level crossing gates after the demolition of the down main line platform, view facing south with the signal box on the far side of the gates. The GNR made several attempts to widen their main line from two to four tracks through Holme but to the end the track layout consisted of up and down main lines and the Ramsey single line. The footbridge, dating from 1936, was removed in 1958. *Author's Collection*

From the outer face of the island platform the Ramsey North branch curved away to the east on a 13 chain radius right-hand curve, descending at 1 in 100 over occupational crossings Nos. 1 and 2 at 69 miles 45 chains, to follow a straight and level course across the black fenlands. Less than half a mile to the north, on the down side of the line were the distinctive silver birches and other deciduous trees of Holme Lode Covert and Middle Covert, part of Holme Fen nature reserve. Because of the importance of free drainage of the fenland soil, the railway company provided nine culverts on this section of line.

At the end of the straight section the branch passed over Whittlesea Mere level crossing No. 13, known locally as Short's crossing, at 70 miles 34 chains from King's Cross, with its adjacent crossing keeper's cottage on the up side of the line, east of the crossing. The branch then eased round a short 40 chains radius right-hand curve to follow a further straight and level course across the edge of Holme Fen. With open fenland on the up side, the trees of Jackson's Covert closed in on the down side of the line before the railway bisected Triangle occupational crossing No. 15, at 70 miles 74 chains, which carried a narrow lane across the rails to New Decoy Farm. The branch penetrated a small copse, known as Railway Covert and then crossed Robinson's accommodation crossing No. 16, at 71 miles 35 chains before curving slightly to the right on a 55 chain radius right-hand curve to Long Drove crossing No. 17 at 71 miles 51 chains, where the B660 road from Holme to Ramsey St Mary's bisected the railway. It was also known to local railwaymen as White's crossing, after the ganger who resided with his wife in the adjacent brick-built gate lodge, located on the up side of the line west of the crossing.

From Long Drove level crossing, protected in both directions by gate distant signals, initially worked from a small ground frame by the crossing keeper but later fixed, the branch straightened and levelled out again to cross the black fenland fields frequently interspersed by drainage ditches and culverts. On the down side of the line the buildings of Chaulderbeach Farm heralded the approach to the only break in the level gradient on the branch after leaving Holme. A short rise and fall at 1 in 66 carried the line over the drainage channel known as New Dyke or Draper's Delph by Nightingale Corner underbridge No. 1, at 72 miles 23½ chains ex-King's Cross, constructed on three cast-iron spans on timber supports. On resuming level track beyond the bridge, the branch curved slightly to the left to continue a straight course over Bunnage's occupational crossing No. 21 at 72 miles 43 chains, which carried an access track to the nearby Willington House. The line then crossed Ugg Mere where a second underbridge No. 2, constructed of timber, carried the railway over Hogmere Drain at 72 miles 62½ chains. St Mary's West occupational crossing No. 22 bisected the line at 72 miles 77 chains before the approach to the entrance points to St Mary's station goods yard, located on the down or north side of the main single line. A few yards further and the branch entered St Mary's station 73 miles 11¼ chains from King's Cross, and 3 miles 62 chains from Holme. The station, which served the village of Ramsey St Mary's, was located half a mile south of the main centre of population near the church of St Mary's. The church suffered from the shifting peaty subsoil and was leaning so dangerously that the spire was later removed. Soon after the opening of the line, a hostelry,

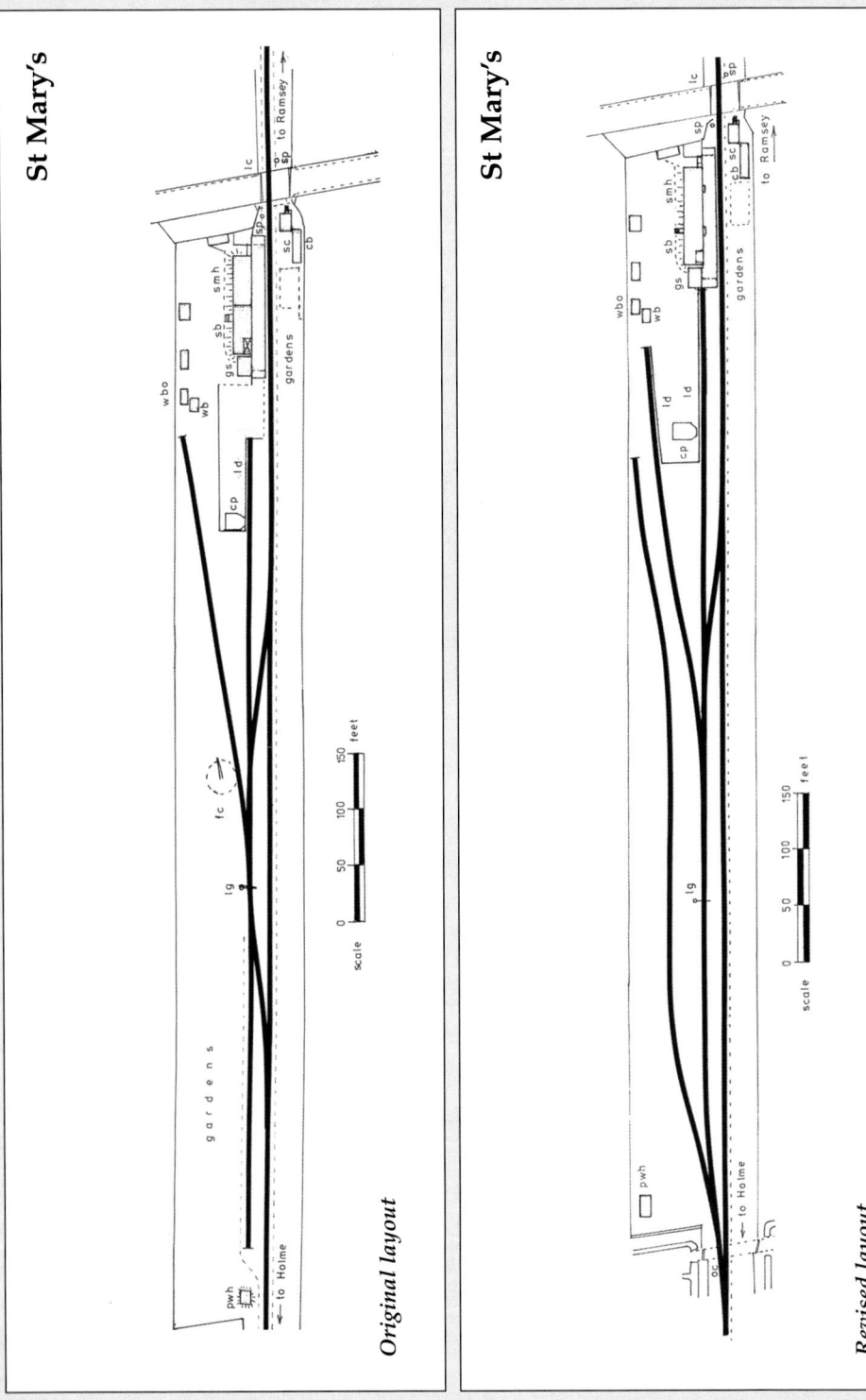

The single platform station at St Mary's facing towards Holme. When this photograph was taken the signals and signal box had been abolished. The goods yard located on the down side of the main single line can be seen beyond the station buildings. *Stations UK*

St Mary's station, 3 miles 62 chains from Holme, looking towards Ramsey North in the 1950s after the demolition of the signal box. Note the wooden sleeper-edged gravel surface platform. The canopy between the station building and the brick transit shed was a later addition to the original structure. The former station master's garden is on the up side of the single line opposite the platform. *Author's Collection*

The exterior and entrance to the station buildings at St Mary's. The steps lead to the booking office whilst the station master's, later foreman-in-charge's, living accommodation is at the far end of the structure. As the building stood on concrete blocks the area underneath became a nesting place for rats in later years. *Author's Collection*

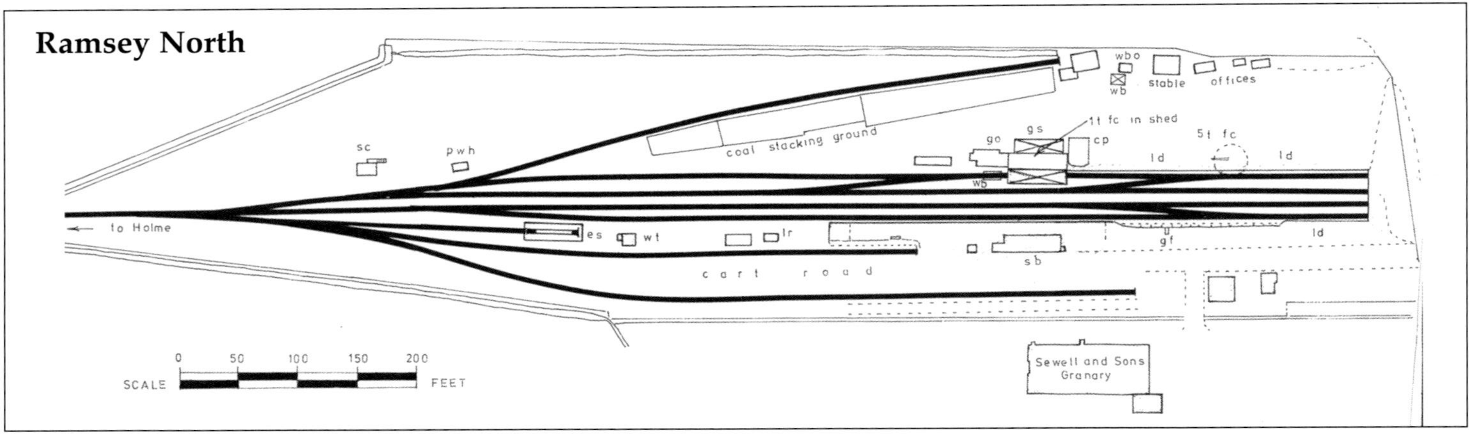

Ramsey North

Ramsey North station from the main single line, showing the points leading to the down side goods yard and up side cart road in June 1963. By this date the granary siding had been removed. *Robert Powell* known as the Railway Inn and owned by John Shelton was established near the station.

The single platform at St Mary's, 150 ft in length on the down or north side of the single line was immediately west of Ugg Mere Covert Road level crossing. The platform was host to the timber station buildings including the residence for the clerk-in-charge, later designated station master, general waiting room, booking office and toilets. The station buildings were supported on concrete piling and the space underneath later became a notorious haunt of rats. The goods yard, served by three sidings, back road 541 feet in length, dock road 304 feet and the 788 ft loop road, was located west of the station and north of the main single line. The brick goods transit shed located at the west end of the passenger platform was built in 1935. Originally a 10 ton capacity fixed crane was provided in the yard but this was later reduced to 3 tons capacity, and by 1912 had been removed.

On the opposite side of the line to the platform was the station master's and later foreman-in-charge's garden. It was here that the wife of the incumbent hung out her washing, always a problem when shunting took place or when trains stopped at the station, as the resultant smuts from the engine chimney could play havoc with the 'whites'.

The distant, home and starter signals in each direction, and level crossing at St Mary's were controlled from the Station signal box, equipped with a 20-lever frame, located on the up side of the line opposite the platform and immediately west of the level crossing. The signalman also operated the entry points leading to the goods yard at the eastern end of the layout but the connection to and from the main single line at the western end was operated from a 3-lever ground frame released by Annett's key, which was kept in the signal box. After the abolition of the signal box on 2nd June, 1936 the Annett's key on the single line Train Staff released the ground frame and points from the single line to the goods yard. At the rear of the signal box an old coach body was utilized as a sack store.

Leaving St Mary's the single branch line bisected Ugg Mere Covert Road level crossing No. 23 at 73 miles 14 chains, and continued a straight and level course for the next 1¼ miles across the black cultivated fields and associated drainage ditches of New Fen, passing in its course Smallholders' occupational crossing No. 26 at 74 miles 02 chains, and Gull's double occupational crossings Nos. 27 and 28 at 74 miles 27 chains.

The main road from Ramsey St Mary's to Ramsey, B1040 ran almost parallel to the branch line and never more than half a mile north of the railway. Five miles from Holme and on the approach to Ramsey, the branch swung to the right on a 44 chains radius curve across the locally titled Poors Land, where several allotments and smallholdings were established close to the railway. On the approach to School Farm Drove level crossing No. 29, at 74 miles 69 chains, the line straightened out to follow a level course to Ramsey North station, 75 miles 09 chains from King's Cross and located on the north-western outskirts of the town adjacent to the gasworks, a few yards short of Station Road.

The terminus was served by a single platform, 240 ft in length on the up or south side of the 575 ft-long loop line. The platform contained the timber station

Ramsey North station 5 miles 65¼ chains from Holme, viewed from the buffer stops. The lines from left to right are the platform road, run-round road, middle road and shed road. In the background is the tank house, which stood adjacent to the former engine shed. In the far distance is the small signal box. The goods shed with double canopies housed a 1 ton capacity crane, whilst another fixed crane of 5 ton capacity stands on the loading dock. In the right background is the coal road. *Author's Collection*

Ramsey North station from the buffer stops at the shunt spur end of the run-round road. To the left is the platform road whilst middle road has been shortened by buffer stops. Points from the run-round road lead to the goods shed and loading dock. *Author's Collection*

Ramsey North station and goods yard in the early 1960s. Beyond the platform, wagons are standing in the 440 feet cart road siding, whilst behind the station buildings is granary road 680 feet long. The goods shed is to the right whilst middle road has been shortened, as has the shed road, and access to the latter is from the run-round loop. *The late Tom Middlemas*

The timber station building at Ramsey North and single platform on the up side of the single line, view facing towards Holme. *Author's Collection*

Ramsey North station facing the buffer stops with the 240 ft-long platform on the up side. Note the lower height of the structure in front of the 'temporary' station buildings. *E. Sawford*

The less than imposing Ramsey North station from the approach road. The goods shed is to the right. *E. Sawford*

Ramsey North station facing towards the buffer stops. Had the original plans been successful, the railway would have continued to St Ives or failing that, to Somersham by joining up with the Ramsey & Somersham Junction Railway line at their Ramsey High Street station, half a mile distant. In the event Ramsey, like its neighbour, became a terminus but retained its through platform.
Author's Collection

Sewell's granary, later owned by Larratt Brothers located to the south of Ramsey North goods yard. The granary had its own siding installed in 1915 and both companies provided considerable traffic to the railway. As late as 1966, ten wagons of bulk grain were dispatched daily to Birkenhead and Manchester.
Author's Collection

buildings including booking office, porters' room, waiting room, ladies' room and gentlemen's toilets. The main single line continued past the platform loop to terminate at the buffer stops, 75 miles 14 chains from King's Cross and 5 miles 65 ¼ chains from Holme. Goods facilities included the 470 ft coal road serving the stacking grounds of Coote & Warren, Fairweather and the Peterborough Co-operative Society, the 780 ft shed road serving the goods shed with its 1 ton capacity crane and loading dock, where the 5 ton capacity fixed crane was located. The adjacent 760 ft-long middle road, which was connected by facing points to the shed road formed a run-round loop. Alongside the shed road in later years was the station master's office, also used by the commercial agent. South of the main single line was the 160 ft former engine shed road, later known as the tank road and the 440 ft cart road, and 680 ft granary or outside road adjacent to Sewell & Sons, later Larratt's granary.

Points and signals at the station were controlled from Ramsey Station signal box, equipped with a 25-lever frame, and located west of the entrance points to the goods yard on the down side of the main single line. A 3-lever small ground frame bolt locked from the signal box was initially provided at the east end of the station to work the points at that end of the station but after signalling was abolished and the signal box closed on 2nd June, 1936 some points were operated from a ground frame and others hand operated.

In addition to the two numbered bridges, the branch also crossed 34 culverts of varying dimensions, provided to allow adequate drainage of the surrounding fenland. The speed limit for trains using the branch was initially 30 mph with a 10 mph restriction round the 13 chains radius curve at Holme. The speed limit was later increased to 40 mph but from 1960 this was reduced to 25 mph because of the deteriorating condition of the permanent way. From the same date trainmen were required to open and close the gates at Whittlesea Mere, St Mary's Fen (Long Drove) and St Mary's level crossings. By the early 1970s the speed limit was again reduced to 15 mph.

It is 75 miles and 14 chains to Kings Cross and 5 miles 65¼ chains to Holme measured from the buffer stops in the foreground at Ramsey North. The view taken in 1957 shows the sparse fenland landscape served by the branch, which was level throughout except for a minor gradient where the line crossed New Dyke on its way the junction at Holme.

Author's Collection

Chapter Six

Permanent Way, Signalling and Staff

The initial permanent way of the Ramsey Railway was formed of flat bottom rails in 23 ft lengths, weighing 65 lb. per yard. The rails were laid on transverse sleepers 9 feet in length by 9 in. by 4½ in., laid three feet apart and two feet apart at the rail joints. The rails were fishplated and secured to the sleepers by fang bolts and coach screws, four fang bolts being used on the outside edge of each length of rail. Both fang bolts and coach screws passed through the flange of the rail. The track was laid on ballast formed of gravel and burnt clay to a depth of 14 inches below the underside of the sleepers. Although land had been purchased for double-track railway and the formation for the most part was 17 ft wide, only a single-track railway was provided. Where sidings and crossing loops were installed the distance between the running lines was six feet and in some places seven feet. When Colonel Yolland inspected the Ramsey Railway on 8th June, 1863, he was dissatisfied with the fixing of the permanent way and requested the railway company to provide additional fang bolts on the inside of each length of rail on the sleeper next to the joints, the work to be completed within three months of the inspection. The inspector also required additional ballasting because of the peaty foundation of the line.

By the early 1880s the initial permanent way had been replaced by 80 lb. per yard bullhead rails in 24 ft lengths, laid on chairs weighing 38 lb., the rails and chairs being fastened to the sleepers by iron spikes and wooden trenails. Around the turn of the century 85 lb. per yard bullhead rails gradually replaced the lighter rails and these sufficed until just before the Grouping. After Grouping the LNER replaced the 24 ft rail lengths with lengths of 30 ft and 45 ft rails, and these became the standard length until after World War II when some 60 ft lengths were employed. Gradual changes were then made and much of the line was relaid using second-hand 87 lb. and 90 lb. rail of GNR and LNER vintage, which had previously served on the main line.

The original gravel and burnt clay ballast was soon replaced by the GNR, who introduced ashes and clinker to the formation. This proved suitable for the light traffic carried and ashes were readily available from the motive power depots on the system. Ashes remained in use until the closure of the line, although on certain sections susceptible to subsidence because of the peaty nature of the subsoil, ballast chippings and packing was added. When supplies of ashes were not available from locomotive sheds, wagon loads were obtained after the 1920s from the sugar processing factories at Peterborough or Spalding.

The maintenance of the permanent way initially appears to have been covered by a permanent way gang at Holme. Later in the 19th century lengthmen were appointed to cover the section from Ramsey to St Mary's exclusive and St Mary's to Holme exclusive, the latter living in the crossing cottage at Long Drove where his wife was employed as resident crossing keeper. Around the turn of the century the maintenance was the sole responsibility of the six-man gang established at Ramsey and remained so until

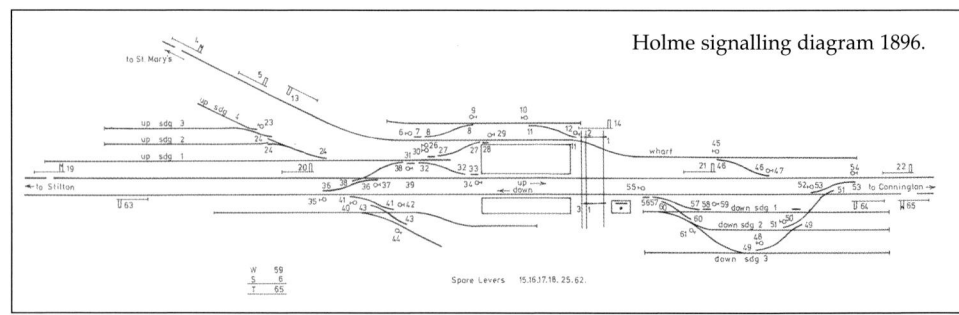

Holme signalling diagram 1896.

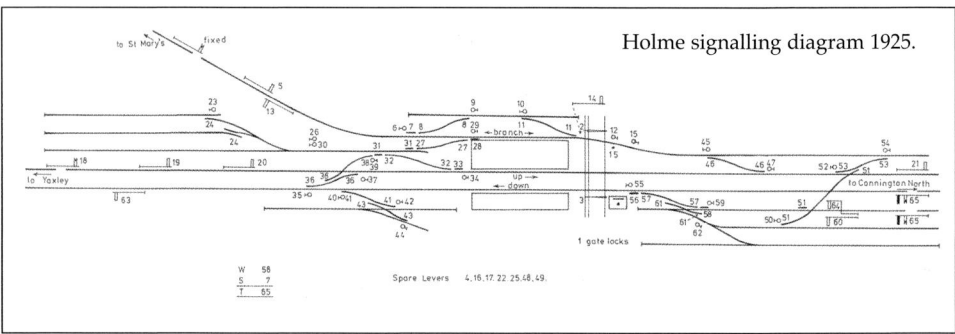

Holme signalling diagram 1925.

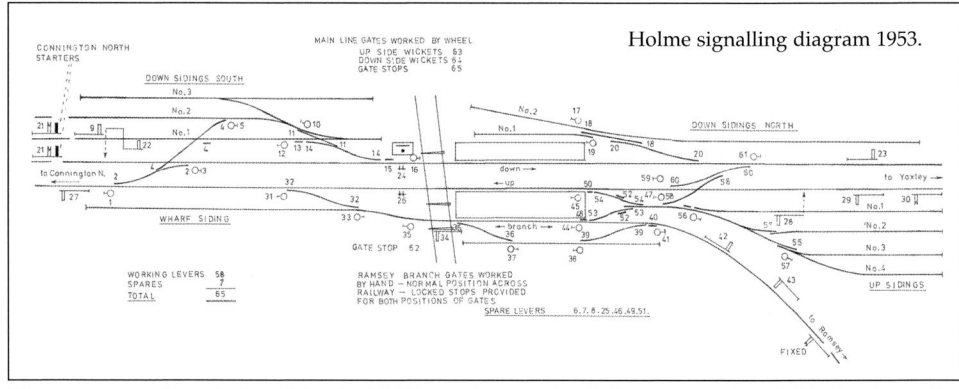

Holme signalling diagram 1953.

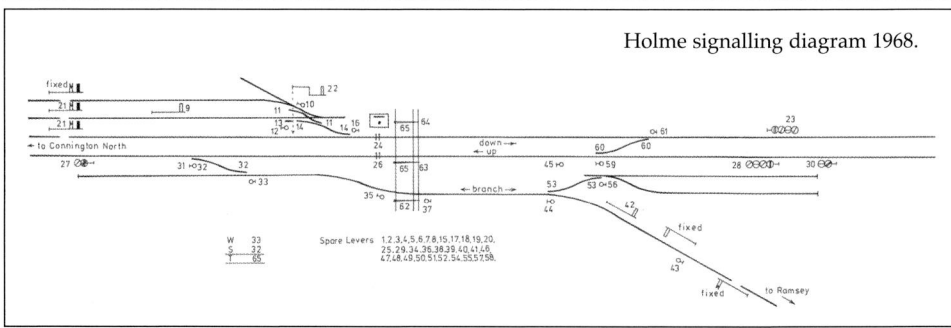

Holme signalling diagram 1968.

the late 1950s. From then until the line closed, maintenance was carried out by staff based at Holme with assistance also provided from the district headquarters at Peterborough.

In addition to attending to the day to day maintenance of the track, the permanent way gang were responsible for cleaning the toilets at the stations where no mains drainage was provided, as well as maintenance of fences and gates. Mr James was one of the longest serving permanent way staff in the area with 47 years as a ganger and permanent way inspector at Holme also covering the Ramsey North branch. Another ganger at Holme was F.W. Mears who died on 27th July, 1928, whilst J.H. Barrick passed away on 16th March, 1937.

Signalling

The initial signalling for the Ramsey Railway conformed to the existing GNR system and was supplied by Stevens & Sons of Southwark, London. Fixed signals for each direction of travel at Holme, St Mary's and Ramsey were mounted on the opposite sides of a single post at each station and displayed three positions. The signal arm positioned at 90 degrees to the post indicated danger, caution at 45 degrees and clear when slotted in the post. At night a revolving spectacle showed red for danger, green for caution and white for line clear. The signals on the approach to the junction at Holme and to the terminus at Ramsey only indicated danger or caution. In addition to the station signals, auxiliary or distant signals were provided 800 yards in rear of and showing the same indication as the fixed signal. The auxiliary signals were to be treated as stop signals but once a train had

Holme signal box was provided with a 65-lever frame to control the points and signals on the Great Northern main line, and connections with the Ramsey branch. The structure on the down side of the line is pictured after the provision of lifting barriers. The signal box was abolished on 2nd June, 1976, when control of the main line signalling was transferred to Peterborough power box. *Author's Collection*

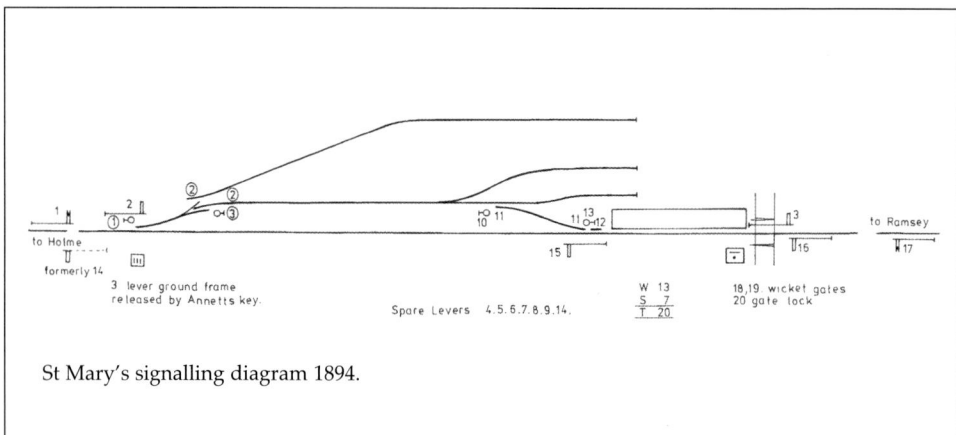

St Mary's signalling diagram 1894.

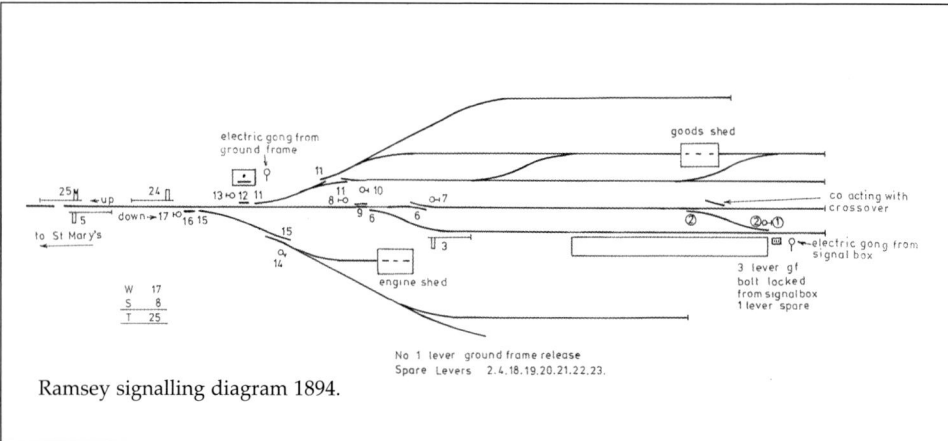

Ramsey signalling diagram 1894.

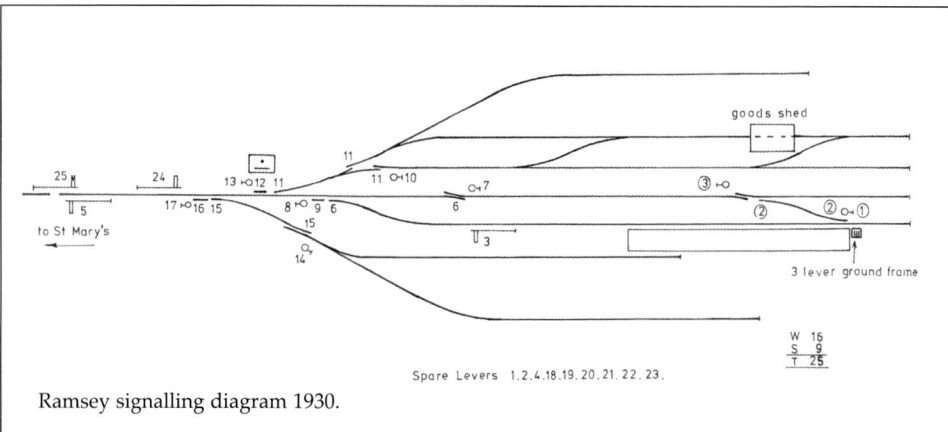

Ramsey signalling diagram 1930.

been brought to a stand the driver was permitted to take his train at caution to the next fixed signal, whilst watching for any possible obstruction on the line. Auxiliary signals were provided at Holme in the up direction, St Mary's in both directions and Ramsey in the down direction.

Traffic on the branch was initially worked on the time interval system, the signals being placed to danger once a train had passed, then placed to caution and then to clear once a prescribed interval had passed. Initially this appeared adequate for the branch was worked on the 'One Train Only' method of operation. Because of increasing traffic the GNR authorities found this method unsatisfactory. In July 1866 the GNR General Manager noted that the Holme to Ramsey line was one of the two lines on the GNR system not equipped for Train Staff and Ticket working. To enable more than one train to use the branch, he recommended alterations and the new method of working was introduced later in the same year utilizing a square-shaped Train Staff lettered 'Holme and Ramsey'. Paper tickets issued in connection with the Train Staff were red in colour for down trains and green for travel in the up direction.

As a result of the Regulation of Railways Act 1889, it was amongst other things mandatory to have block telegraph signalling of trains on all lines except those worked by 'One Engine in Steam' or Train Staff without Ticket method of working. On investigation, and with the full agreement of the owning company the GER, the GNR decided the volume of traffic on the Ramsey branch was insufficient to warrant the expense of the equipment. To save costs and rationalize within the scope of the Act, 'One Engine in Steam' (or two or more coupled together) using the existing square-shaped Train Staff replaced the existing Train Staff and Ticket method of working. The only exception to this rule was when a second locomotive entered the branch to clear a failed train or assist in the event of an accident. The same Act also required the interlocking of points and signals on running lines but as both main line companies were deeply involved with the equipping of more important lines, no immediate alterations were made on the Ramsey branch.

Unfortunately for the owning GER and the leasing GNR, the collision of a light engine with coaching stock standing at the platform at Ramsey on 15th November, 1893, highlighted the lack of progress. Ten passengers were slightly injured in the incident and at the subsequent Board of Trade inquiry the inspecting officer was highly critical of the delay in providing the interlocking between the points and signals, especially as the work should have been completed by 20th November, 1892. The inspector required the GER, as the owning company to carry out the work immediately to comply with the requirements of the 1889 Act. The GER and GNR authorities conferred as to the required work and Mr Johnson, the GNR signal & telegraph engineer, provided estimates. At the GER Board meeting on 20th February, 1894, William Birt, the General Manager advised that interlocking of signals and points on the branch would cost £2,008. The GER Engineer considered the estimates reasonable and the Directors readily agreed to the GNR executing the work, with costs borne by the GER subject to the terms of the lease. After receiving the contract in March at a cost of £1,518 7s. 7d., McKenzie & Holland completed the new signal boxes at St Mary's and Ramsey, and associated signalling by October 1894. New somersault signals replaced the old semaphore signals on the branch. These had been developed as a result of the Abbots Ripton disaster of 21st January, 1876, when snow clogged up signals and wires preventing the arms of the old type semaphore signals returning to danger. The GNR later adopted a centrally balanced somersault arm, which was proof against clogging snow. These signals with pitch pine posts and cedar arms had cast- and wrought-iron fittings. The fishtail distant signals were painted the same red as stop signals and showed the same red and green aspects to drivers at night. This anomaly remained until the LNER assumed responsibility for the line when the distant signals were gradually repainted the familiar yellow with black > thereon.

The first signal box at Holme was provided in 1859 but by 1875 a new structure had been provided which in 1881 had a 42-lever Saxby & Farmer 5 inch rocker frame with 37 working and five spare levers. This signal box was inadequate for the additional installations and new signalling was authorized in 1895. As an interim measure six additional levers were added to the existing frame making a total of 48 levers. The new signal box provided at Holme in 1896 contained a 65-lever Saxby & Farmer 4 in. rocker frame, of which a total of 10 levers worked signals, 15 points, 26 ground disc signals, six locking bars, two wicket gates and one the level crossing locking, the other five, later six, being spares.

At St Mary's the Station signal box installed in 1894 contained a 20-lever McKenzie & Holland cam and tappet frame with 4 in. centres. Of the working levers seven controlled signals, two ground discs, one a pair of points, one a locking bar, one the crossing gate lock and two the wicket gate locks, whilst six levers were spare. A 3-lever McKenzie & Holland cam and tappet ground frame with 4 in. centres, released by Annett's key normally kept in the signal box, operated points leading from the main single line to the goods yard at the Holme end of the station. Later the frame had 13 working and seven spare levers when the up advance starting signal was abolished.

Ramsey Station signal box contained a 25-lever McKenzie & Holland cam and tappet frame with 4 in. centres, with four levers operating signals, six ground disc signals, three points, three locking bars and one a bolt lock, the other eight were spares. A 3-lever McKenzie & Holland cam and tappet ground frame with 4 in. centres and bolt locked from the signal box was provided to operate points at the east end of the run-round loop near the buffer stops. By 1930 the frame had 16 working and nine spare levers.

On the branch, Holme station was protected by a down starting signal, located 252 yards from the signal box and up distant, outer home and inner home signals, the latter at 1,036 yards and 322 yards from the signal box, together with an up starting signal protecting the branch level crossing. St Mary's station was protected by down distant signal, 1,046 yards, the down home, 296 yards and starting signals, 3 yards from the signal box. In the up direction the distant, home, starting and advance starting signals were located 763 yards, 25 yards, 86 yards and 410 yards from the signal box respectively. Ramsey was provided with an operating, later fixed distant signal, 722 yards, and home signal, 98 yards from the signal box in the down direction and starting and advance starting signals in the up direction, 125 yards and 298 yards from the signal box respectively. By 1925 Holme signal box had 58 working and seven spare levers and this number continued through the next four decades although from September 1954 a new GN1 type frame replaced the original, but by 1970 with rationalization the box had 33 working and 32 spare levers. It was then reported to be a 65-lever Railway Signal Co. frame. Holme signal box was abolished on 2nd June, 1976 and replaced by new crossing box controlling the level crossing but under the supervision of Peterborough power signal box.

Holme signal box, which controlled all signals and points on the main line was open continuously, whilst St Mary's and Ramsey signal boxes were open for all trains running on the branch. In 1897 and during GNR days when there was fog or falling snow, fog signalmen manned Holme up branch distant and Ramsey down distant signals.

Rationalization of branch operating was achieved as early as 2nd June, 1936 after the withdrawal of the majority of passenger services, when the signal boxes at St Mary's and Ramsey were abolished. Most of the existing points were converted to hand operation, although points leading to and from the main single line were operated by ground frames, released by Annett's key attached to the Train Staff. In the interests of economy a one-lever economical locking frame was installed at St Mary's allowing the shunter to release the Jack Catch locking bar and change the points to the goods yard with one pull. At the same time all signals at St Mary's and the terminus were removed, except for the distant signals, which remained *in situ* to remind drivers of the level crossings and also on the approach to Ramsey North station. Later these distant signals were replaced by sighting boards. At Holme all the branch signals were retained. When the branch finally closed to freight traffic in 1973, the ex-GNR down starter and up home signals at Holme were purchased by the Peterborough Locomotive Society and were used as the down starting signals at Wansford station on the Nene Valley Railway. A GNR signal on the spare road at Holme was also purchased by a former signalman and erected in the garden of a house near the station.

It is not generally known that the railways first brought standard time to villages and towns throughout the country. After the opening of the branch, the clocks on the stations at St Mary's and Ramsey provided Greenwich Mean Time to the local community. As late as 1881, King's Cross station telegraph office transmitted, by single needle telegraph daily at 10.00 am, the time-signal to signal boxes and stations along the main line. Sandy South signal box passed the message on to Holme signal box, where the signalman then passed the message on by telegraph to St Mary's and Ramsey, so that clocks and watches could be adjusted and corrected. Local villagers and townsfolk then adjusted their timepieces to correspond with the station clocks. Although post office and telegraph offices and later the wireless took over the same function, the railway 10.00 am time signal continued to be transmitted until the early 1960s.

Traffic staff

Motive power and civil engineering staff, are dealt with in their respective sections but a brief mention must also be made of traffic staff serving on the branch. As far as can be ascertained the GNR appointed the initial staff for the Ramsey Railway from existing servants when the line opened to traffic, and Nathan Jackson was appointed station master at St Mary's with John Allen at Ramsey. Allen proved to be the longest serving station master on the branch, serving at the terminus for 26 years before retiring in 1889. George Vincent replaced Allen. Jackson was obviously seeking promotion for he only remained at St Mary's until the mid-1870s when William Perkins was transferred to the position. The station master at Holme during the initial years was George Gregory who stayed at the junction station until 1877. By 1890 Arthur Joseph Potts had taken over as station master at Holme, whilst John Drury was at St Mary's. In 1895 Thomas Lawrence took over from George Vincent as station master at Ramsey and in 1898 James Webb Eyre succeeded Drury at St Mary's. Eyre had started his railway career at Luton but stayed for only a short while on the branch before gaining promotion as station master at Etwall on the Derby to Burton-on-Trent line. By this time Potts had moved on from Holme and William Pickworth became station master but his tenancy was short, for he retired in the early 1900s and was replaced by Herbert Bonner. Bonner remained as station master at Holme for nearly 20 years. By 1906 Arthur George Dodd had replaced Lawrence as station master at Ramsey and served for a period of eight years at the terminus. In 1904 Herbert E. Cullen had been appointed to the post of station master at St Mary's but his stay was short for he was transferred away in 1910 and was replaced by John William Palmer. Cullen must have liked the area, for in the same year he returned to the branch, succeeding Dodd as station master at Ramsey. J.W. Palmer remained for only a few years at St Mary's and by 1914 Walter Edwin Green had taken over. H. Gayton, who was promoted from the position of chief goods clerk at Biggleswade to St Mary's, replaced Green. His tenure was short for he soon transferred to Tempsford on the main line. Gayton finally retired in March 1932 from the position of station master at Clacton-on-Sea.

When the LNER assumed control of the line, Cullen was station master at Ramsey and as part of rationalization of branch working he initially took over control of Ramsey High Street (later Ramsey East) station as well as St Mary's. However, after a short period this arrangement was found to be inconvenient and Ramsey East reverted to the responsibility of the station master at Somersham. Cullen was succeeded by J.H.L. Davies, who had previously served at Warboys and at Burwell on the ex-GER Mildenhall branch before being appointed to Ramsey North. In March 1928 he was promoted to Market Rasen and was presented with a small desk clock by staff at Ramsey North and St Mary's before transferring. Davies subsequently retired in December 1929. J.H. Sexton, who was in charge at Holme, was promoted to station master at Huntingdon in February 1930 and was presented with an eight day clock by the station staff, and a newspaper and letter rack by the staff at St Mary's. He subsequently retired in 1943 after 47 years' service. Sexton was replaced by F. Pickworth, son of the above mentioned William Pickworth, who transferred from Wainfleet in May 1930. After the withdrawal of the major proportion of passenger services from the line in February 1931, the post of station master at Ramsey North was withdrawn and the branch came under the jurisdiction of the station master at Holme. Pickworth remained at Holme and in charge of the branch stations until February 1940 when he was promoted to take charge at Huntingdon, being presented with a Westminster chiming clock on behalf of the staff before transfer. H. Soanes, station master at nearby Yaxley was transferred to acting station master at Holme in September 1940. He remained at the junction station until January 1944 when L. Walker, station master at Chartley on the ex-GNR Uttoxeter to Stafford branch was transferred on promotion to take charge at Holme. By December 1945 Walker was on the move again, this time to Alford Town but the resultant vacancy was not filled until April 1946 when A.F. Fear, station master at Little Bytham was promoted to Holme. He had commenced his railway career with the GER at Somersham and was later a clerk at Guyhirne before gaining promotion to station master at Coldham on the March to Wisbech branch and then Little Bytham. The post of station master at Ramsey North was reinstated in February 1940 when J. Benton was promoted to the post from the goods office at Huntingdon North. He retired on 11th June, 1949 after 44 years' service, a date coinciding with the withdrawal of the post. A.F. Fear served at Holme until April 1962 and at a retirement gathering was presented with a set of bowls woods by staff, Mrs Fear receiving a bouquet of carnations. After the closure of Holme station the administration of the branch was covered by the area manager at Huntingdon, with J.C. Holmes in charge when the branch closed in July 1973.

Until 1931 when the majority of passenger train services were withdrawn, a guard was on the establishment at Ramsey North working a regular middle turn, relieving a Holme guard and later New England or Peterborough guard who worked the early turn. At the end of his turn a New England or Peterborough guard relieved the Ramsey guard, who worked the last up train on the branch. After 1931 New England or Peterborough guards worked all passenger and freight services across the branch. By 1947 when passenger services were finally withdrawn from the Ramsey North line the staff establishment at the branch stations was:

> **Holme** - Station master, 2 booking clerks, 2 foremen, 3 porters, 2 goods porters, 3 signalmen, a relief signalman, 7 gate lads and a motor driver. Three of the gate lads were responsible for opening and closing the level crossing gates between the hours of 7.00 am and 10.00 pm. At other times the signalmen opened and closed the gates. The other four lads' duties included general assistance around the station on passenger, parcels and goods work, and cleaning and trimming signal and gate lamps at Holme, Connington and the Ramsey branch stations. W. Morris was foreman at Holme before passing away on 9th December, 1927, whilst Walter Merrishaw retired as signalman on 31st May, 1934. On retirement station master Pickworth presented Merrishaw with a Westminster chime clock at a gathering of staff. Another signalman at the junction station W.T. Mills retired on 5th August, 1937. For a number of years a goods guard was on the establishment at Holme and A. Rollings was the last holder of the post. He died on 18th November, 1939. After World War II, the Hills family, father and son were signalmen at Holme, whilst the daughter was the booking clerk.

Staff at Ramsey North in 1967. From left to right are Eric Marshall, foreman, Yvonne Endersby, the last goods clerk to serve at the station and Ron Sharpe, a Peterborough guard.
Author's Collection

Class '08' diesel electric shunting locomotive No D3445 waits at Ramsey North during shunting operations in 1967. Beside the driver is Peterborough guard Ron Sharpe and foreman Eric Marshall.
Author's Collection

Jack Shaw, who served as foreman at St Mary's for almost 30 years, checks the time as the branch freight train waits to depart from the intermediate station for Ramsey North behind a 'J4' class 0-6-0. *Author's Collection*

St Mary's - Foreman, porter/signalman and goods porter/shunter. One of the porter's posts was later withdrawn and the station became an unstaffed public siding from 2nd May, 1960. Jimmy Moore was the last porter/signalman and Jimmy Carter the last goods porter/shunter to serve at this station. In the 1890s John Sutton supervised the collection of agricultural products from the fen districts on behalf of the GNR. The traffic was worked by water into Chatteris Dock on the GN&GE Joint line as well as St Mary's, Holme and Peterborough. He was familiarly known as 'The Commodore of the GNR Fleet' and retired from the position of goods agent at Peterborough on 21st December, 1927.

Ramsey North - Station master, booking clerk, 2 goods clerks, foreman, senior checker, road vehicle driver and porter/signalman. The booking clerk was withdrawn after the cessation of the passenger services, although for some years previously the duties had included a large proportion of goods accounts. A guard's position was also on the establishment and J. Jeffrey was the last occupant of the post before he retired. He passed away on 12th September, 1935. In July 1946 P.H.D. Bowles a clerk at Barton-on-Humber was transferred to the goods office to be followed in February 1947 by H.T.C. Sharman a clerk from the Stamford East goods office. Working foreman Stanley Eustace retired on 23rd March, 1947 after 40 years' service, 26 of which had been spent at Ramsey North. At a gathering of staff and well-wishers he was presented with a silver cigarette lighter and cigarette case on behalf of the staff by station master Benton, whilst J.R. Major, Chairman of Ramsey UDC presented him with a cheque on behalf of local farmers and merchants. When the post of station master was withdrawn in 1949, a commercial agent was appointed to take charge for a short period. Over the years as traffic declined other posts were withdrawn, the last goods clerk leaving in 1967. Before the line closed to traffic the only position on the branch was that of the foreman at Ramsey North, who also controlled the sidings at St Mary's as part of his duties. Eric Marshall spent almost 25 years at Holme and on the branch, latterly as the penultimate goods foreman at Ramsey North.

Chapter Seven

Timetables and Traffic

With the building of the railway to Ramsey, the promoters were led to believe the line would save the locality from economic stagnation and lead to an increase in the population. Although Ramsey was a relatively large town for the area, to the east and west lay flat and dank fenland, broken only by drainage waterways and dykes. The failure of the railway to extend beyond the town to St Ives or Somersham, and later to connect with the younger Ramsey & Somersham Junction Railway, sealed the Ramsey branch to a domiciled fate. The original company and its successors certainly provided a faster transit for commodities to local and London markets and eased the problems of landowners and farmers, but the line brought little increase in population and the expected expansion never materialized.

The uncertainty as to whether Ramsey owed trading and market affinity to the county town of Huntingdon or the neighbouring city of Peterborough, was a problem for the railway companies, and the GNR, as operators for the GER, always worked the line as a self contained unit, providing a through service to Peterborough only on infrequent occasions. As most of the population of the area was employed on the land, little use was made of the line for daily travel and longer journeys were often deterred by the poor connections maintained with main line services at Holme. The 5¾ mile branch, however, boasted a service of up to seven passenger trains weekdays only, for a catchment area with a population of around 6,000, as the undermentioned figures show.

	1861	1871	1881	1891	1901	1911	1921	1931	1951
Ramsey	4,500	4,734	4,617	4,684	4,823	5,328	5,135	5,180	5,770
St Mary's	1,088	1,128	1,224	1,241	1,245	1,340	1,284	1,240	1,200
Holme	644	655	629	627	658	648	575	559	531
Total	6,232	6,517	6,470	6,552	6,726	7,316	6,994	6,979	7,501
Total excl. Holme	5,588	5,862	5,841	5,925	6,068	6,668	6,419	6,420	6,970

Holme certainly provided little local traffic to the branch, whilst from 1889 the Ramsey High Street to Somersham branch provided the townsfolk of Ramsey with an alternative route to London and East Anglia. Journey times to the capital by this route were, however, longer with more changes of train and the Ramsey North route tended to be preferred by those travelling to London.

The initial service provided by the GNR in 1863 of six trains in each direction on weekdays only departed Ramsey at 8.05, 9.05 am, 12.45, 2.25, 3.40 and 6.24 pm with an equal number of return workings from Holme. This service was evidently over generous and from 1st August, 1863 the number of trains was reduced to five by the withdrawal of the 2.25 pm ex-Ramsey and its return working, whilst the 9.05 am departure was brought forward to 9.00 am and the 6.24 pm left one minute later. Trains called at St Mary's by request only. By 1864 the weekdays-only passenger train service remained at five in each direction with an additional train each way on Saturdays.

Down					SO		
		am	am	pm	pm	pm	pm
Holme	dep.	8.30	10.00	1.50	3.00	4.05	7.30
St Mary's	dep.	*	*	*	*	*	*
Ramsey	arr.	8.45	10.15	2.05	3.15	4.20	7.45

Up					SO		
		am	am	pm	pm	pm	pm
Ramsey	dep.	8.05	9.00	12.40	2.20	3.40	6.20
St Mary's	dep.	*	*	*	*	*	*
Holme	arr.	8.25	9.15	12.55	2.45	3.45	6.35

* Trains call at St Mary's by request only. SO – Saturdays only.

The fastest journey time between London and Ramsey was available by the 12 noon train from King's Cross, which connected with the 1.50 pm ex-Holme giving a 2 hours 5 minutes timing. In the up direction passengers departing Ramsey on the 6.20 pm were afforded a two hour timing to King's Cross with one change at Holme. Some of the connecting times at Holme left much to be desired, for passengers catching the 12.50 pm up train from Peterborough waited 50 minutes at the junction for the 1.50 pm to Ramsey.

In 1873 the GNR passenger timetable showed three passenger and two mixed trains in the down direction and two passenger and three mixed trains from Ramsey. Down trains departed Holme at 10.20 am, 1.25, 2.25, 4.45 and 7.10 pm, whilst up trains departed from Ramsey at 9.00 am, 12.30, 1.55, 3.25 and 5.40 pm. Trains ran on weekdays only and called at St Mary's by request when there were passengers to pick up or set down, on notice being given to the guard at the previous station. Passenger trains were allowed 15 minutes and mixed trains 18 to 20 minutes for the 5¾ mile journey. The fastest journey time between London and Ramsey had deteriorated and was available to passengers on the 4.55 pm ex-King's Cross which connected with the 7.10 pm mixed train from Holme to give a 2 hour 35 minute timing to Ramsey. In contrast the slowest journey was only 15 minutes longer in duration by catching the 10.50 am from King's Cross. In the up direction King's Cross could be reached in 2 hours 20 minutes on the connection off the 12.30 pm ex-Ramsey, whilst passengers catching the 5.40 pm mixed train arrived in London three hours later, as they had to wait 57 minutes for a connecting southbound train at Holme.

The 1877 working timetable showed an increased service of five passenger, two mixed and one goods train in each direction, weekdays only. As the branch engine was based at Ramsey shed, the first working was an up mixed train at 8.05 am. The second mixed train departed Ramsey at 12.35 pm whilst the goods train ran at 4.25 pm, both making mandatory calls at St Mary's. In the down direction the mixed trains departed Holme at 10.20 am and 1.21 pm, the former making a mandatory stop at St Mary's, as did the afternoon goods which departed the junction at 5.15 pm. Passenger trains were allowed 18 minutes and mixed trains 25 minutes for the journey from Holme to Ramsey, whilst all

Table 27.]   RAMSEY AND HOLME LINE.

Passenger Fares from & to London.				Miles from Holme		TO RAMSEY.					TO RAMSEY.—WEEK DAYS ONLY.				
RETURN.		SINGLE.						morn	morn	after.	after.	after.	after.	after.	after.
1st cls.	2nd cls.	1st cls.	2nd cls.	3rd cls.										4 20	
s. d.	s. d.	s. d.	s. d.	s. d.		LONDON (King's Cross) .... dep.		7 30	10 30	...	...	...	...	6 0	...
...	...	...	...	...	...	Huntingdon .... ,,		9 56	1 8	...	2 16	4 30	...	6 10	...
...	...	...	...	...	...	Holme, from South .... ,, arr.		10 14	1 21	...	2 26	4 44	...	6 57	...
...	...	...	...	...	...	Peterboro' .... dep.		1	12 55	...	...	...	...	...	...
...	...	...	...	...	...	Holme, from North .... arr.		9	12 40	...	...	...	...	6 57	...
16 0	13 6	8 8	...	...	3¼	HOLME .... dep.		1020	1 5	30	4 45		7 5	
17 8	13 6	10 3	...	0¼	5	St. Mary's .... ,,		*	*	*	*		*	
20 0	14 0	10 6	...	2½	6	RAMSEY .... arr.		1035	1 40	4	5 5		7 25	
								morn.	after.	after.	after.		after.	

Horses, if the same Property.			Dogs.	Miles from Ramsey.		FROM RAMSEY.				FROM RAMSEY.—WEEK DAYS ONLY.				
One.	Two.	Three.					morn	after.	after.	after.	after.	after.	after.
s. d.	s. d.	s. d.	s. d.				9 0	12 25	2 5		3 25	6 10	6 35
19 6	29 0	38 6	1 6		RAMSEY .... dep.		*	*	*		*	*	*
17 6	25 0	32 6	1 6	2¼	St. Mary's .... ,,		9 15	12 45	2 20		8 43	6 5	6 37
				6	HOLME .... arr.								
...	...	...	...	1¼	Holme, for North .... dep.		10 14	1 31	...	5 48	...	...	...
...	...	...	...		Peterboro' .... arr.		10 27	1 51	...	4 0	...	...	...
...	...	...	...	16	Holme, for South .... dep.		9 27	12 48	2 30	...	...	6 57	...
...	...	...	...		Huntingdon .... ,,		9 44	1 5	2 40	...	...	7 17	...
...	...	...	...	74¼	LONDON (King's Cross) .... arr.		11 50	3 5	5 40	...	...	8 35	...
							after.		after.			after.	

No Sunday Trains.

Trains stop at St. Mary's by Signal only when there are Passengers to take up or set down. Passengers signing to or from etc.

GNR public timetable 1874.

trains, other than those specifically mentioned, only called at St Mary's when there were passengers to pick up or set down and on notice being given to the guard at Holme or Ramsey or by signal stop at the intermediate station. In effect all passenger trains on the branch could run as mixed as the timetable specified that mixed trains were limited to not more than 10 vehicles exclusive of coaching stock, whilst passenger trains could take a tail load of not more than six wagons of cattle.

In 1885 before the closure of Ramsey engine shed, the branch was supporting a weekdays-only service of six passenger, one mixed and two goods trains in each direction. On Wednesdays and Saturdays an additional passenger train departed Ramsey at 7.15 am with a non-stop run to Holme, where the balancing return working was a goods train which departed the junction at 7.50 am, serving St Mary's from 8.02 to 8.09 am, before arriving at Ramsey at 8.16 am. The 2.40 pm goods ex-Holme and the return 4.50 pm goods from Ramsey were through trains from and to Peterborough worked by a New England depot engine and men. All other workings on the line were covered by the Ramsey-based locomotive. Where more than 30 wagons required movement from the branch, the 4.50 pm goods conveyed only traffic for destinations north of Holme. Traffic for stations south of the junction were conveyed on additional goods trains, which ran only as required, departing Ramsey at 8.05 pm and returning from Holme at 9.15 pm. Passenger trains continued calling at St Mary's by request only and all were permitted to convey not exceeding six wagons of cattle. Mixed trains on the branch were restricted to 16 wagons in addition to the passenger carrying vehicles.

By July 1895, the GNR provided a down service of five Saturdays-excepted (SX), six Saturdays-only (SO) passenger, two mixed and two goods trains with two balancing light engine movements, the line being worked by a New England locomotive and men throughout the day. The early morning goods train ran from New England yard to Peterborough and on to Holme before departing at 5.58 am for Ramsey. The mixed trains on the branch, which ran at 7.43 am and 1.14 pm ex-Holme were augmented by the 4.30 pm ex-Holme which was permitted to take cattle wagons if fitted with brake pipes. The additional SO passenger train departed the junction at 2.26 pm. There were light engine trip workings from St Mary's to Ramsey after shunting the yard and at 5.37 pm ex-Holme, after working the 4.50 pm goods train from Ramsey, so the locomotive could work the 5.55 pm up passenger train. The 7.50 pm goods train ran non-stop from Holme to Ramsey. In the up direction six passenger trains ran SX with seven on Saturdays augmented by one mixed working. Two goods trains handled freight traffic whilst a light engine ran from Ramsey to St Mary's to perform shunting at the intermediate station. An additional goods train ran if required but if no traffic was to hand the engine ran light to Holme, en route to New England shed. The single mixed train of the day was the first up working at 7.05 am ex-Ramsey and ran as such to clear essential overnight freight and perishable traffic from the terminus. The SO additional passenger train departed Ramsey at 2.05 pm, whilst the light engine to St Mary's at 2.55 pm only ran when necessary to shunt traffic in the yard. Both up goods train workings at 4.50 pm and 7.00 pm collected traffic from St Mary's, as did the 9.00

RAMSEY AND HOLME BRANCH—NO SUNDAY TRAINS.
SINGLE LINE. TRAIN STAFF STATIONS:—HOLME AND RAMSEY.

Miles from Holme.	DOWN.	1 Gds. C	2 Mxd.	3 Pass.	4 Pass.	5 Mxd.	6 Pass. Sats. only.	7 Lgt. engn. B	8 Pass. A	9 Lgt. engn.	10 Pass.	11 Gds.	12 Pass.					
		a.m.	a.m.	a.m.	a.m.	p.m.	p.m.	p.m.	p.m.	p.m.	p.m.	p.m.	p.m.					
...	HOLME...dep.	5 58	7 43	9 28	10 42	1 14	2 45	...	4 10	5 37	6 35	7 46	8 34					
3¼	St. Mary's {arr. dep.	6 10 6 22	7 54	9 38	10 51	1 25	2 50	3 30 3 35	4 19 4 24	... 5 50	6 44 6 49	... 8 0	8 43 8 48					
5¼	RAMSEY...arr.	6 30	7 59	9 41	10 56	1 30												

Miles from Ramsey.	UP	1 Mxd.	2 Pass.	3 Pass.	5 Pass.	6 Pass. Sats. only.	7 Lgt. engn. B	8 Pass.	9 Goods	10 Pass.	11 Gds.	13 Pass.	14 Lgt. Eng. D	15 Gds. E
		a.m.	a.m.	a.m.	p.m.	p.m.	p.m.	p.m.	p.m.	p.m.	p.m.	p.m.	p.m.	p.m.
	RAMSEY...dep.	7 5	9 10	9 45	12 40	2 15	2 54 3 0	3 40	4 36	5 55	7 0	8 10	9 0	9 6
2	St. Mary's {arr. dep.	7 12 7 24	9 15 9 24	9 50 9 58	12 46 12 54	2 20 2 29	...	3 45 3 54	4 41 6 5 5 17	6 0 6 9	7 6 7 26 7 38	8 15 8 24	... 9 15	9 20 9 32
5¼	HOLME...arr.													

A May take cattle if wagons are fitted with brake pipes. See special instructions.
B To shunt St. Mary's yard when necessary.
C From New England, see train 83, page 54. Will not run when 15 is running.
D To New England, see train 644, page 35. Will not run when 14 is running. To New England, see train 644, page 35.
E When necessary.

GNR working timetable 1897.

pm goods train, which ran if required through to New England yard, Peterborough. In the absence of traffic the branch locomotive returned light to New England shed at 9.00 pm, clearing Holme at 9.15 pm. Passenger trains in both directions were permitted 13 to 16 minutes running time between Holme and Ramsey, whilst mixed trains were given 16 to 19 minute timings. Goods trains with intermediate stops of between 12 and 20 minutes at St Mary's took 32 to 36 minutes other than Saturdays.

Five years later the working timetable for July 1900 showed little alteration. The basic down service consisted of five passenger, two SX and three SO mixed trains, augmented by two goods workings and two light engine movements. The locomotive allocated to work the line brought the early morning goods train to the branch from New England and departed Holme at 5.58 am. Other trains showed only minor adjustments in timing to coincide with changes to the main line timetable, whilst the additional SO train departing Holme at 2.36 pm was downgraded to 'mixed', although it still retained a passenger train timing of 14 minutes for the journey to Ramsey. The only other significant change was the later running of the last down passenger train at 9.15 pm ex-Holme. The light engine returning from St Mary's yard to Ramsey at 3.36 pm only ran if required, whilst the 4.10 pm Holme to Ramsey passenger train was permitted to take cattle wagons if fitted with brake pipes. In the up direction the GNR provided six SX and seven SO passenger and one mixed trains. A basic service of two goods trains was augmented by an additional up train at 9.35 pm ex-Ramsey if there was traffic to clear from the branch stations. As in earlier years, if there was no traffic the engine was worked light to New England shed departing Ramsey in the same path. In contrast to the additional SO down working, the additional up SO train continued to run as a passenger service departing Ramsey at 2.15 pm. The other light engine movement was the conditional 2.55 pm to St Mary's, where the locomotive shunted the yard for 30 minutes before returning to the terminus.

The service provided on the Ramsey branch remained basically the same for the next few years and except for minor adjustments in timings the down trains in 1905 were unaltered. In the up direction, however, the 'as required' additional goods train was replaced by an additional passenger train which departed Ramsey on Wednesdays and Saturdays only at 9.42 pm. After stabling the coaching stock at Holme the engine returned light to New England depot. The coaching stock was worked back to Ramsey attached to the 5.58 am goods train on the Thursday or the following Monday morning. It is interesting to note the discrepancy in the timetable, for the light engine was booked to depart the terminus two minutes before it was due to arrive at Ramsey with the last down passenger train, the 9.23 pm ex-Holme.

The working timetable for 1910 showed a down service of five SX and six SO passenger trains and two SX and three SO mixed trains with two goods and one light engine working. The morning goods, departing Holme at 5.58 am was a through working from New England yard and brought the branch locomotive to the line. The attractions of Huntingdon and Peterborough for weekend shopping excursions engendered the provision of two additional SO passenger trains, departing Holme at 3.47 and 4.18 pm. The operation of these services called for smart locomotive and station working at both Ramsey and Holme with four minutes allowed for uncoupling, running round the train and attaching again before departure. This included the collection of additional cattle wagons at Holme between 4.14 and 4.18 pm, before setting off for Ramsey. Here in the outback of GER territory, the GNR was possibly pre-empting on Saturday afternoons the famed procedures for rapid arrivals and departures, which were later to become the hallmark of the 'Jazz' suburban services from platforms 1, 2 and 3 at Liverpool Street. On SX the late afternoon passenger train departed Holme at 4.09 pm and was permitted to take cattle wagons if fitted with brake pipes. The only other goods working of the day was the 7.56 pm ex-Holme which ran non stop to Ramsey, whilst a notable omission was the 'Q', as and when required, light engine working from Ramsey to St Mary's and back to shunt St Mary's yard, which was discontinued. In the up direction the service consisted of six SX and eight SO passenger trains, augmented by the one early morning mixed train which departed Ramsey at 7.00 am. In addition to the 2.10 pm SO departure, another passenger train ran at 4.04 pm from Ramsey to Holme omitting the St Mary's stop, although the same 14 minutes timing was allowed. The final up passenger train departed at the earlier time of 8.15 pm from Ramsey, with the light engine returning to New England shed at 9.35 pm after hauling the 9.15 down passenger train. Two goods trains served the branch at 4.35 SX, 4.40 SO and 7.05 pm, the former being allowed 24 minutes and the latter 20 minutes to shunt the yard at St Mary's. Passenger train timings across the branch continued to vary between 13 and 14 minutes, whilst mixed trains were permitted 16 to 19 minutes.

The working timetable for 1916 showed five passenger SX, six passenger SO, two mixed and two goods trains plus one light engine movement in the down direction and six passenger SX, eight passenger SO, one mixed and two goods trains augmented by one light engine SX in the up direction. The 5.58 am goods train from Holme was a through train from Peterborough, as before. Of the passenger trains the 4.09 SX and 4.18 SO were permitted to take cattle if the wagons were fitted with brake pipes, whilst the additional Saturday trains departed Holme at 3.47 and 4.18 pm, and returning from Ramsey at 4.04 and 9.45 pm, the latter being formed of engine and carriage brake only. On SX the engine returned light from Ramsey at 9.35 pm en route to New England shed. The 4.35 SX and 4.40 pm SO up goods trains booked to call at St Mary's from 4.41 to 5.05 pm and 4.46 to 5.10 pm respectively were allowed slightly more time to clear the yard of wagons but shunting was to cease at 5.20 to enable the train to depart by 5.25 pm. This was to allow the engine time to run back light from Holme to Ramsey to work the 6.00 pm Ramsey to Holme passenger train without causing delay. Any outstanding wagons not taken by the goods train had to be cleared by the 7.05 pm goods train ex-Ramsey, which was allowed 20 minutes at St Mary's for shunting duties.

By 1921 four passenger, three mixed and one goods train operated in the down direction, with six passenger, one mixed and one goods train in the up direction. There were two paths for light engine movements in both directions. The 5.58 am goods train ex-Holme was a through working from Peterborough, whilst the down light engine paths were at 3.05 SO and 5.37 SX. The 4.15 pm ex-

Holme passenger train was allowed to take cattle traffic provided the wagons were fitted with brake pipes. In the up direction the goods trains departed Ramsey at 2.15 SO and 4.45 pm SX, with both shunting at St Mary's to pick up wagon loads for onward transit. The SX working was timed to stay at St Mary's from 4.51 to 5.15 pm but in exceptional circumstances a further five minutes was allowed with the proviso that the train was not to depart later than 5.25 pm to enable the engine, having stabled the wagons, to return promptly to Ramsey to work the 6.15 pm up passenger train. At the end of the day after working the last down passenger train ex-Holme at 7.57, the engine returned light at 8.15 pm to Holme SX en route to Peterborough but if there was sufficient wagons awaiting dispatch a goods train was substituted departing Ramsey at 8.20 pm.

The final service provided by the GNR from October 1922 showed four passenger, three mixed, one goods and one light engine working in the down direction. The engine booked to work the branch continued to arrive with the goods ex-New England yard and departed from Holme at 5.58 am, whilst the three mixed trains ran during the morning and early afternoon. An unbalanced working meant that after hauling the up goods train the branch engine returned light at 3.05 SO and 5.37 pm SX from Holme to work the next passenger service. The 4.15 pm passenger train ex-Holme continued to provide a service for farmers and was permitted to work cattle if the wagons were fitted with brake pipes. A radical alteration after the war years was the much earlier departure of the last down passenger train at 7.45 pm ex-Holme. In the up direction the timetable provided for six passenger, one mixed and one goods train. The mixed train was the first departure from Ramsey at 6.55 am to clear overnight traffic, whilst the SO goods train departed at 2.15 pm. On SX the goods train left Ramsey at 4.45 pm with 24 minutes allowed to shunt St Mary's yard. Although booked to depart at 5.15 pm from the intermediate station, provision was made for shunting operations to continue for a further five minutes, with departure put back to 5.25 pm at the latest, as described above. In the event of outstanding freight traffic not being cleared from St Mary's or Ramsey by the afternoon goods train, the branch engine worked a special freight train at 8.20 pm from Ramsey to Holme. If no goods train was warranted, the locomotive returned light to New England shed from Ramsey at 8.15 pm.

From 1923 the LNER offered a similar basic service but in the spring of 1926 the miners' dispute, which later led to the General Strike in May, brought a cessation of services. The curtailment was short lived, for trains were then re-introduced worked by volunteer crews who, because of the absence of railway staff, were required to open and close the level crossing gates on the branch. After a week of abnormality, railwaymen returned to duty but the continuing coal strike caused problems for the railway companies and the LNER management decided to reduce train services on all lines from 31st May, 1926. Passenger trains departed Ramsey North at 6.55, 9.05, 9.48 am, 3.40 and 6.15 pm returning from Holme at 7.43, 9.30 and 10.15 am, whilst the 4.15 was altered to depart at 4.40pm on Saturdays only and the 7.50 to 7.35 pm on Saturdays only. This service was only temporary and when conditions allowed, the full service was reinstated.

In 1927 the LNER increased the service by one additional train each way on Saturdays. In the down direction this ran as a mixed train departing Holme at

RAMSEY NORTH AND HOLME BRANCH. NO SUNDAY TRAINS.
SINGLE LINE. TRAIN STAFF STATIONS. HOLME AND RAMSEY NORTH.

Distance from Holme	DOWN. Class	1 Gds. D.	2 Mixd.	3 Pass.	4 Mixd.	5 Mixd. SO	6	7	8 Light Eng. SO	9	10 Pass. G	11 Pass. SO	12 Lgt. Eng. S	13 Pass.	14 Pass.	15	16	17	18
M. C.		a.m.	a.m.	a.m.	a.m.	p.m.			p.m.		p.m.	p.m.	p.m.	p.m.	p.m.				
0	HOLME........dep.	5 58	7 43	9 30	10 15	1 35			3 5		4 15	5 10	5 37	6 38	7 50				
3 62	St. Mary's {arr.	6 10	7 54	......	10 24	1 44			......		......	......	......	6 47	7 59				
	St. Mary's {dep.	6 22	7 54	9 38	10 27	1 48			......		4 23	5 19	......	6 47	7 59				
5 65½	RAMSEY NORTH..arr.	6 30	7 59	9 43	10 33	1 54			3 18		4 29	5 24	5 50	6 52	8 4				

Distance from Ramsey	UP. Class	1 Mixd.	2 Pass.	3 Pass.	4 Pass.	5	6 Gds. D. SO	7	8 Pass.	9 SO	10 Pass. Gds. D. S	11	12 Pass.	13 Pass.	14 Lgt. Eng.	15 Gds. D. SR	16 Pass. SO	17	18
M. C.		a.m.	a.m.	a.m.	p.m.		p.m.		p.m.	p.m.	p.m. E		p.m.	p.m.	p.m. F	p.m.	p.m.		
0	RAMSEY NORTH...dep.	6 55	9 5	9 48	12 45		2 15		3 40	4 41	4 45		6 15	7 15	8	8 15	8 30		
2 24	St. Mary's {arr.	7 2	9 10	......	12 50		2 21		......	......	4 51		......	......	......	8 41	8 35		
	St. Mary's {dep.	7 2	9 10	9 53	12 50		2 45		3 45	4 46	5 15		6 20	7 20	......	8 51	8N45		
5 65½	HOLME........arr.	7 14	9 19	10 2	12 59		2 57		3 54	4 55	5 27		6 29	7 29	8 30	9 1	8N55		

E Shunting operations at St Mary's must cease at 5.20 p.m. and the train must be despatched by 5.25 p.m. at latest. If necessary 15 up must be run to clear out any wagons left off 10 up. F Runs when 15 up does not run. G May take cattle if wagons are fitted with brake pipes. N Light engine. Coaches to be worked back to Ramsey North on Monday mornings.

LNER working timetable for the branch 1928.

1.40 pm, whilst the last down passenger again left the junction at 7.50 pm. In the up direction only a few variations were made to the timing of services. On Mondays to Fridays the branch engine was booked to return to New England shed running light at 8.14 pm ex-Ramsey North, or if traffic required, hauling an additional goods train to clear the line of essential wagon loads and departing the terminus at 8.35 pm. On SO, however, the locomotive hauled the additional 8.30 pm Ramsey North to St Mary's passenger train and, after stabling the coaching stock in the yard proceeded light to New England at 8.45 pm, clearing Holme 10 minutes later. The coaching stock was worked back to Ramsey North the following Monday morning attached to the 5.58 am goods ex-Holme. This peculiar working persisted for the next two years.

The final full timetable on the Ramsey North branch in 1930 showed a down service of four SX and five SO passenger, three mixed, one goods and one light engine working. The morning goods train, by now the 4.55 am from Peterborough Spital yard, was worked by the branch engine and departed from Holme at 5.58 am. All the mixed trains ran in the morning or early afternoon, with the last departing Holme at 1.35 pm SX and five minutes later SO. The SO light engine working ran at 3.05 pm after bringing the goods up from Ramsey North, but on SX it ran at the later time of 5.37 pm from Holme. The 4.15 pm passenger train was permitted to take cattle wagons if the vehicles were fitted with brake pipes, whilst the LNER at long last provided a through passenger train between Peterborough and Ramsey North on Saturday evenings, departing the cathedral city at 10.40 and Holme 11.01 pm, to cater for people attending the theatre or cinema. In the up direction a service of six SX, seven SO passenger and one mixed train was provided, augmented by one goods and one light engine working. The mixed train cleared overnight traffic from the branch departing Ramsey North at 6.55 am whilst the SO goods ran at 2.15 pm. On SX freight was cleared from the branch by the 4.45 pm goods train, which if required was permitted an additional 5 minutes at St Mary's before departing at 5.25 pm to maintain main line connections at Holme. If additional traffic required to be cleared SX, the light engine working from Ramsey North at 8.15 was cancelled and an additional goods ran at 8.35 pm to clear the branch yards of wagons. On Saturdays after working the 7.50 pm down passenger train from Holme, the branch engine worked the empty coaching stock back to Peterborough, departing Ramsey at 9.20 to form the 10.46 pm Peterborough to Ramsey North train. This entailed a later light engine working back to New England shed at 11.21 pm from Ramsey North and involved an additional New England crew relieving the late turn men at Peterborough.

The drastic reduction of services from 2nd February, 1931 left only three passenger trains in each direction on the branch, with none after the 10.15 am Holme to Ramsey North train. Buses were then left to cater for intending patrons, leaving the line clear for freight traffic. By 1936 the down service consisted of one passenger, two mixed and three goods trains with two passenger, one mixed and three goods workings on the up road. The branch engine provided by New England shed arrived on the line with the goods from Spital yard and hauled the 5.58 am goods train ex-Holme. After three round trips with passenger or mixed trains, culminating in the 10.15 am ex-Holme

HOLME AND RAMSEY NORTH BRANCH. NO SUNDAY TRAINS.

SINGLE LINE (ONE ENGINE IN STEAM). TRAIN STAFF STATIONS. HOLME AND RAMSEY NORTH.

Distance from Holme	DOWN. Class	1 Gds. D. a.m.	2 Mixd a.m.	3 Pass. a.m.	4 Mixd a.m.	5 Gds. D F p.m.	6 Gds. D p.m.
...	HOLME ... dep.	6 58	7 43	9 12	10 15	1 40	4 30
1	St Mary's { arr.	6 10	...	10 24	1 52	...	...
1	St Mary's { dep.	6 22	7 54	9 39	10 27	1 59	...
5	RAMSEY NORTH ... arr.	6 30	7 59	9 44	10 33	2 7	4 35

	UP Class	Distance from Ramsey N'th	1 Mixd. a.m.	2 Pass. a.m.	3 Pass. a.m.	4 Gds. D a.m.	5 Gds. D J p.m.	6 Gds. D K p.m.
	RAMSEY NORTH ... dep.	0	6 55	...	9 5	9 49	11 25	5 40
	St Mary's { arr.		...	...	9 10	9 54	...	8 46
	St Mary's { dep.	3½	7 2	9 19	...	...	11 40	6 10
	HOLME ... arr.	6½	7 14					6 22

F. To shunt at Ramsey North as required and return light or with goods as 5 up.
J. Shunt Holme Yards and return as 6 down. K. Engine to work 597 down except on Saturdays when it works as required.
Y. Engine to shunt Ramsey North Yard and return as 4 up.

LNER working timetable 1937.

mixed train with arrival at the terminus at 10.33 am, the branch engine shunted Ramsey North yard before departing with the 11.25 am non-stop goods to Holme. Two further goods trip workings at 1.55 and 4.10 pm ex-Holme and 2.40 and 5.00 pm from Ramsey North were interspersed with yard shunting at both Holme and Ramsey North. St Mary's goods yard was only served by the 5.58 am and 1.35 pm freight trains in the down direction and the 5.00 pm ex-Ramsey North on the up road. After working the latter train, the branch locomotive returned light to Peterborough.

Soon after the outbreak of World War II the LNER introduced an emergency timetable with effect from 2nd October, 1939 and the Ramsey North branch service was reduced to one passenger train in each direction departing Ramsey North at 8.55 and returning from Holme at 9.40 am. A footnote to the timetable advised that an omnibus service commencing after 12.00 noon was provided by Eastern Counties Omnibus Co. between Ramsey, St Mary's and Holme to connect with main line services. After the immediate concern of invasion had passed the emergency service was aborted and for the remainder of the early years of hostility the advertised passenger train service provided departures from Ramsey North at 6.50 and 8.55 am returning from Holme at 7.35 and 9.52 am. Additional unadvertised services were run for servicemen going to or leaving neighbouring airfields, whilst at other times the passenger coaches allocated to the branch were attached to the freight trains to provide the necessary accommodation as required. By 1944 the meagre passenger service offered by the LNER remained at two morning trains in each direction on weekdays only. Up services departed Ramsey North at 7.00 and 8.05 am returning from Holme at 7.35 and 9.55 am, the trains being allowed between 12 and 20 minutes for the journey.

Immediately after hostilities had ceased the LNER introduced the only advertised Sunday services to run on the branch. The May 1946 timetable showed two passenger, one empty coaching stock working on Fridays only, three goods SX and two goods SO in the down direction, whilst on the up road two passenger trains ran Fridays excepted (FX) and three Fridays only (FO), two goods SO and three SX were also provided. The first down goods was the 4.00 am Yaxley to Ramsey North, which departed Holme at 5.58 am. After arriving at the terminus the locomotive was utilized to work the two passenger services each way together with yard shunting at Holme before arriving back at Ramsey North at 10.27 am. After further shunting at the terminus the engine then worked the 10.45 am non-stop goods train to Holme clearing the branch at 11.00 am, although at most times this train ran early. On Fridays the empty coaching stock ran from Peterborough, departing at 10.57 to Ramsey North arriving at 11.58 am, the train being propelled by the locomotive from Holme to the terminus. The stock formed the 12.32 through train to Peterborough North arriving at 1.04 pm, which ran non-stop across the branch and was provided essentially for servicemen from local airfields taking weekend leave. After the passenger train had cleared the branch the afternoon weekdays-only goods departed Holme at 1.40 pm and called at St Mary's to set down wagons before arriving at Ramsey North at 2.07 pm. Once shunting was completed the return working departed at 2.35 pm with a 15 minute timing non-stop to Holme. On

LNER poster advertising the withdrawal of passenger train services from the Holme to Ramsey North branch on and from Monday 6th October, 1947. *Michael Brooks*

Saturdays excepted a further goods train served the branch at 4.15 returning at 5.00 pm from Ramsey North and calling at St Mary's in order to clear the yard of revenue earning traffic. The sole Sunday train which ran for servicemen returning to local airfields was the 10.28 pm Peterborough North to Ramsey North through working, which after reversing at Holme from 10.42 to 10.50 pm, deposited its passengers at the darkened terminus at 11.03 pm. As the station was closed and unstaffed on Sundays, the locomotive was unable to run-round the train and the return empty coaching stock was propelled by the engine back to Holme departing Ramsey North at 11.13 and clearing the branch at 11.31 pm.

After nationalization the service in 1952 showed a reduction to two class 'K' freight trains in each direction. The initial down working departed Holme at 7.50 am and after 10 minutes allowed for shunting at St Mary's arrived at Ramsey North at 8.40 am. With an hour allowed for shunting duties the up train departed the terminus at 9.40 calling at St Mary's only if required, and arriving back at Holme at 10.30 am. After yard duties the locomotive returned light to New England shed on Mondays to Fridays whilst on Saturdays it worked a further round trip to Ramsey North departing Holme at 11.15 am and returning from Ramsey North at 12.45 pm. Both workings called at St Mary's. On SX the second down freight train departed Holme at 4.15 pm and ran non-stop to Ramsey North with 25 minutes permitted for the journey. On completion of shunting and forming of stock, the up train departed at 5.00 pm and called at St Mary's from 5.06 to 5.30 pm, clearing the branch of all revenue earning wagons and arriving at Holme at 6.02 pm to connect with the main line pick-up goods. On arrival with the 12.40 pm SO up goods train the branch engine remained at Holme working to Control instructions.

By 1963 the decline of railborne freight had resulted in the provision of only one freight train in each direction across the branch. On Saturdays the down train departed Holme at 10.30 am with a 30 minute timing to Ramsey North and calling at St Mary's only if required. The train returned from Ramsey North at 12.15 pm with 15 minutes allowed for shunting at St Mary's before clearing the branch at Holme at 12.58 pm. On Mondays to Fridays the train ran at the later time of 1.30 pm and called at St Mary's from 1.52 to 2.01 pm with an arrival at Ramsey North eight minutes later. Considerable time was permitted at the terminus for shunting and sorting of wagons before the train started back at 4.30 pm, with a generous 62 minute timing to the junction which included 24 minutes for clearing traffic from St Mary's. As the timings were excessive for the short journey the train often arrived before time at Holme, although care was taken that all wagons containing revenue-earning traffic were cleared from the station yards. Unfortunately the decline of freight traffic continued and by 1967 the SO service was withdrawn. From then on the SX down branch goods departed Holme at 13.30 and was booked to call at St Mary's for nine minutes before arriving at Ramsey North at 14.09 hours. Exports from St Mary's yard were now few and far between, and so the 16.40 up goods from Ramsey North called only if required, before arriving at Holme at 17.05. By 1972 the train departed Holme at 08.35 with a 35 minutes timing to Ramsey North arriving at 09.10. The return trip departed at 09.35 again with a 35 minutes timing to the junction because of the 15 mph speed limit imposed because of the deteriorating condition of the track and the need for the train crew to open and close the intermediate level crossing gates. This timetable of one SX train in each direction remained unaltered until the line closed completely on and from 2nd July, 1973.

Fares

During the first years of operation, first, second, third and Parliamentary fares were offered. The latter signified third class travel at 1*d*. per mile, which had to be provided on every line by at least one train in each direction on weekdays under an Act of 1844. The full fare structure to stations on the line in 1875 was:

King's Cross to	Single			Return		
	1st	2nd	3rd	1st	2nd	3rd
Holme	10s. 9d.	8s. 3d.	5s. 9d.	21s. 0d.	16s. 6d.	11s. 6d.
St Mary's	11s. 9d.	9s. 0d.	6s. 0½d.	22s. 6d.	17s. 3d.	12s. 1d.
Ramsey	12s. 3d.	9s. 3d.	6s. 2½d.	23s. 6d.	17s. 6d.	12s. 5d.

By 1904 fares from London, King's Cross to Ramsey had been reduced to first single 11*s*. 0*d*., first return 21*s*. 1*d*.; third single 5*s*. 11*d*., third return 11*s*. 10*d*.

In 1911 market tickets were issued at the branch stations for return journeys to Peterborough on Wednesdays and Saturdays, the third class return fares being:

Ramsey to Peterborough	1s. 4d.
St Mary's to Peterborough	1s. 2d.
Holme to Peterborough	9d.

By 1937 fares to the branch stations from King's Cross were:

	Holme	St Mary's	Ramsey North
First single	14s. 6d.	15s. 4d.	15s. 9d.
Third single	8s. 9d.	9s. 3d.	9s. 6d
First monthly return	17s. 9d.	18s. 9d.	19s. 3d.
Third monthly return	11s. 9d.	12s. 6d.	12s. 9d.
First ordinary return	29s. 0d.	30s. 8d.	31s. 6d.
Third Ordinary Return	17s. 6d.	18s. 6d.	19s. 0d.

Local fares were:

Holme to	1st single	3rd single	1st return	3rd return
St Mary's	10d.	6d.	1s. 0d.	9d.
Ramsey North	1s. 3d.	9d.	1s. 6d.	1s. 0d.

Just prior to the withdrawal of passenger traffic from the branch in 1947 the following fare structure was in operation:

King's Cross to	1st single	3rd single	1st monthly return	3rd monthly return
Holme	20s. 4d.	12s. 3d.	24s. 8d.	16s. 5d.
St Mary's	21s. 3d.	12s. 10d.	25s. 9d.	17s. 3d.
Ramsey North	22s. 0d.	13s. 4d.	26s. 11d.	17s. 11d.

After the withdrawal of passenger services from the line Ramsey North station staff continued to issue paper tickets to various destinations from Peterborough and Huntingdon at rates applicable from those stations.

Excursions

From the opening of the line to passenger traffic the GNR, and later the LNER offered the inhabitants of Ramsey, St Mary's and Holme excursion facilities to London, seaside resorts and other places of interest. The programme, however, were not as generous as those enjoyed by many other GNR branches. Excursions were not aimed specifically at pleasure seekers but were also offered to local growers and farmers. Such excursions included cheap return fares to the annual Smithfield Show held in London in December, and various horticultural shows during the summer months.

Five days after the opening of the line, Monday 27th July, 1863, the combined Sunday Schools of the Ramsey area travelled to Peterborough. The train loaded with 50 adults and over 300 children departed Ramsey at 12.45 pm 'amidst the liveliest demonstrations of interest on the part of the general public' and in complete contrast to the opening day. The train made leisurely progress along the branch before reversing at Holme and took an hour to reach the cathedral city. Here the children were assembled at the recreation ground before consuming a 'bountiful supply of buns'. Games of football and cricket followed, before the group visited the cathedral and then 'consumed more buns'. The party departed from Peterborough by the 6.30 pm train but considerable time was spent at Holme waiting the branch connection so that Ramsey was not reached until 8.00 pm.

In the ensuing months cheap day excursion fares were offered to London at 3s. 6d. for the Odd Fellows Fete at Crystal Palace, which attracted over 100 passengers from Ramsey but a later trip for the Foresters Fete was priced at 4s. 0d. Peterborough was always a popular destination and cheap day returns were issued for the brass band concert when over 250 people travelled. For the annual fair held in October the demand was so great that the train carried 450 passengers from Ramsey, and by the time another 100 or so had joined at St Mary's, some were unable to find room in the carriages, and were conveyed in open goods wagons attached at the rear of the formation. On 6th December, 1864, residents of St Mary's and Ramsey were offered an excursion to London and so attractive was the 4s. 0d. fare for the 140 mile round trip that over 100 passengers made the journey.

Excursion facilities continued to be offered on a regular basis and in July 1866 the cheap day return tickets were available from the branch stations to Huntingdon for the races on the 26th and over 200 tickets were issued at Ramsey.

Although London and Peterborough were the favoured destinations for excursion fares from Ramsey, by the 1880s outings to the seaside were becoming increasingly popular, with the GNR offering regular trips to Skegness. Typically in September 1881, an excursion left Ramsey at 8.20 am so that townsfolk could enjoy the delights of the Lincolnshire coastal resort. After arrival at 10.35 am passengers were able to enjoy a full day before returning at 5.50 pm.

The LNER maintained the programme of excursions and cheap day facilities after Grouping and even after the withdrawal of the major proportion of passenger services in 1931. Empty coaches were worked to Ramsey North and after passengers had joined the train at the branch stations, the stock was returned to Holme to be coupled to excursion trains on the main line bound for such destinations as Skegness, Bridlington, Cleethorpes, Lincoln, York and Scarborough. On the outbreak of World War II excursion facilities were withdrawn but after the hostilities in the late 1940s and 1950s occasional trains still ran to Skegness together with school excursions to York.

Goods traffic

In the years prior to the opening of the Ramsey Railway a considerable number of carriers served the town, travelling with their horse-drawn waggons over poorly-made roads. S. Peppers travelled from Ramsey to Chatteris on Fridays whilst John Cullip, William Colbert and S. Peppers journeyed to Huntingdon on Saturdays. W. Ingle travelled to Peterborough on Saturdays and John Chapman provided a service to Whittlesea on Mondays, Wednesdays, Fridays and Saturdays. Henry Lant also served that town on Mondays, Wednesdays and Fridays travelling from the Jolly Sailor public house at Ramsey. He was later appointed ECR carrier but in October 1858 local railway management found he was not paying over the correct percentage dues to the company, and his contract was withdrawn. The new contract was awarded to Mr Watts. Three carriers also served St Ives on Mondays, John Cullip, W. Ingle and William Colbert. After the opening of the Ramsey line the carriers' services gradually reduced, until by the 1880s they were only providing a feeder service to and from the local railway goods yards.

The Ramsey Railway was conceived and built essentially to alleviate the economic stagnation of the area, partly brought about by the agricultural depression and partly by isolation from the main railway system. The new line provided an effective transport outlet for farmers and growers in the area and the initial freight traffic handled soon confirmed the optimistic forecast of the promoters for barley, wheat, hay, straw, vegetables and coal were quickly transferred from the carriers' carts to the railway. So great was the impact that fen lighter traffic using New Dyke to offload at the wharf by Holme station reduced considerably. The wharf, however, continued to handle a small amount of outgoing vegetable traffic to the railway and incoming coal traffic from the railway until 1947.

By 1869 freight traffic from the branch stations was being dispatched to places as far afield as Boston, Hitchin, Wakefield, Leeds and Lincoln as well as the London destinations of King's Cross, Blackfriars, Mint Street and Poplar. Consignments to the local markets at Huntingdon and Peterborough were the highest tonnages dispatched from St Mary's and Ramsey, although rather disappointingly some of the carrier services continued to convey produce.

Milk, loaded in the familiar 17 gallons churns, was regularly sent from both stations on the branch to Holme for connection with trains to Huntingdon or Peterborough. Two loads were normally dispatched during the summer months, with only one consignment in the winter by the mid-morning train. The area was not noted for dairy farming and therefore the relatively small amounts of traffic conveyed by the branch trains was lost to the railway when the major portion of passenger services were withdrawn in 1931. From then on road carriers collected the churns from the farms and conveyed the produce direct to Huntingdon or Peterborough.

Livestock handled at the branch stations was two-way traffic with young calves imported from all parts of the country including Ireland and Scotland. The fattened cattle were later exported to Huntingdon, Peterborough, Bedford and the London cattle markets. The busiest day for the transhipment of cattle on the branch was Saturdays when animals were loaded at Ramsey, St Mary's and Holme for conveyance to Huntingdon and Peterborough livestock markets. To cater for urgent goods and cattle traffic in the 1870s, passenger trains on the branch were permitted to convey not exceeding one vehicle fitted with brake pipes attached behind the rear brake or carriage brake van. Return workings on market days brought imports of sheep and pigs as well as cattle purchased by local farmers. Ramsey held a market every Wednesday although it was a mere shadow of the larger markets held at St Ives, Huntingdon and Peterborough. The busiest time for the movement of livestock on the branch coincided with the annual cattle fair held at Ramsey each year on 22nd July (Sundays excepted), when additional cattle trains were run before and after the event. Similar trains also ran on the Ramsey High Street branch.

A small amount of horse traffic was also conveyed and although not on the same scale as cattle traffic, the local gentry often made use of horseboxes to convey their steeds to local hunt meetings, shows or bloodstock sales. After World War I horsebox traffic declined, cattle wagons, however, were regularly in use on the branch until after World War II.

From the opening of the line, coal traffic was handled at Holme, St Mary's and Ramsey, usually in privately owned wagons from Clipstone, Kirkby, Stanton, Shirebrook and other Midlands and Yorkshire collieries. The wagons travelled via Peterborough where New England yard acted as a clearing house for loaded wagons en route to the branch and for empty wagons returning to the collieries. As early as August 1863 John Jenkins of Huntingdon was advertising supplies of coal from St Mary's and Ramsey goods yards via his agent Negus Meace of Ramsey. Prices ranged from 17s. 6d. per ton for Durham coal to 13s. 0d. per ton for Darlington soft coal. Coal was also delivered by rail in large quantities to Ramsey gas works, located alongside the station, which pre-dated the railway. By the early 1920s coke was also conveyed for horticultural purposes but after World War II much of this traffic was conveyed by road. Local coal and coke merchants who handled traffic at St Mary's and Ramsey included Coote & Warren, Peterborough Co-operative Society and J. Fairweather. Coal was also conveyed to the Royal Air Force base at Upwood.

For a short period after World War I the Molassine Co. Ltd of Rito Works, Stanground, Peterborough, manufacturers of animal feed, established a peat works at Speechly's Farm on Wood Walton Fen. A 2 ft gauge tramway was contructed in 1918 to assist with the transport of peat to the works. The line ran south into Wood Walton Fen for about 1½ miles extending beyond Goodliff's Lodge Farm and there were several short sidings along the route. After delivery in the tramway wagons at the works, the peat was shredded by a stationary steam engine before being taken by the tramway to the nearby wharf on the navigable Burbridge Stream, for onward transportation by barge to either Holme or St Mary's stations. Here, the GNR forwarded the consignments to various destinations but the traffic lasted for only a few years before the project was abandoned because of the poor quality of the peat.

Other imports handled at the branch stations included seed potatoes from Scotland, fertilizer and farm implements and machinery. Balancing outgoing traffic included potatoes, carrots, swedes, turnips, parsnips, mangold wurzels and other root crops to London markets at Covent Garden and Spitalfields. A considerable tonnage of celery was also dispatched from St Mary's. As late as 1972 seed potato traffic was still conveyed to local farmers. In later years, especially after World War II, wagon loads of scrap metal were sent from St Mary's and Ramsey North to Scunthorpe. The chief outgoing traffic from Holme was grain and potatoes, the main imports in return being coal, coke and seed potatoes. Unusual incoming traffic to Holme goods yard during the 1940s and 1950s were old and damaged Rolls Royce cars, which were renovated and restored by a local enthusiast.

In the late 1920s and 1930s many of the fenland roads remained unmetalled – dust tracks in summer or muddy morasses in wet weather. County councils undertook a rolling programme of road improvements, which involved levelling the surface before covering with granite chippings and tarmacadam. Much of this material was delivered by rail to the branch stations from where the material was offloaded and taken to site by horse and waggon. The granite and tarmacadam was then levelled by steam roller. As late as 1972 hardcore was still being delivered by rail to Ramsey North.

Sugar beet was extensively cultivated in the area of Ramsey, St Mary's and Holme from the mid-1920s. The months from October to January were busy times in the goods yards as farm carts, tumbrels and primitive road lorries brought the beet to be loaded into rail wagons for onward conveyance to the sugar processing factories at Woodston (Peterborough) and Spalding. Often wagon labels were manipulated to ensure certain farmers' consignments of beet were given priority journeys to the factories. For this operation, station staff were offered the price of a pint or invited to the local hostelry. After World War II much of the sugar beet was transferred to road haulage but an increase came as a result of the closure of the Ramsey East goods yard (September 1956) when farmers transferred their produce to Ramsey North for onward conveyance.

Seed and grain traffic were important commodities handled at Ramsey and over the years Messrs Sewell's received and sent many tons of traffic by rail especially after their siding was installed in 1915. The business was later taken over by Larratt Brothers who, as late as 1966 were regularly dispatching 10 loaded bulk grain wagons daily to Birkenhead and Manchester. Fertilizer traffic was also an important earner for the branch and as early as 1866 Evison and Armitage opened a works manufacturing artificial manure at Ramsey, importing raw material and dispatching the finished product by rail. In 1888 the annual output was 2,000 tons, with 1,000 tons of sulphuric acid as a by-product.

Initially the GNR employed independent carriers to provide a cartage service to and from the branch stations but after World War I the company was providing their own service at Holme and Ramsey.

During World War II meat and other foodstuff was conveyed across the branch for storage at the Ministry of Food cold storage depots in the area. Armaments and bombs for local airfields and aviation fuel, notably RAF Upwood were offloaded at Ramsey North and conveyed to site by road during the hours of darkness. To assist with this traffic an additional siding was installed in 1940. Ammunition trains were formed of sheeted open wagons bearing red flashed labels instructing staff to 'Shunt with great care' and 'Place as far as possible from engine, brake van and wagons labelled inflammable'.

The following facilities were available for handling goods and livestock traffic at the branch stations:

Holme	Loading dock
	Loading gauge
	Fixed crane 1 ton 10 cwt (later 1 ton) capacity
	Cart weighbridge
	Weighing machine
	Goods shed
	Cattle dock
	Cattle pens
	Stable
St Mary's	Loading dock
	Loading gauge
	Fixed crane 10 tons (later reduced to 3 tons) capacity, removed by 1912
	Cart weighbridge
	Weighing machine
	Goods shed
	Cattle dock
	Cattle pens
Ramsey North	Loading dock
	Loading gauge
	Fixed crane 5 tons capacity
	Fixed crane 1 ton capacity, in goods shed
	Cart weighbridge
	Weighing machine
	Goods shed
	Cattle dock
	Cattle pens
	Wagon weighbridge
	Stables

The GNR Appendix to the working timetable showed the following loads for freight traffic on the branch:

1877 – Hawthorn's or Sharp's 4-wheel coupled tender locomotives
Holme to Ramsey }
Ramsey to Holme } Maximum load goods and coal, including brake van 30 wagons

1881 – Large 4-wheel coupled tender locomotives
General and pick-up goods loaded   29 wagons
Coal, coke, stone and other minerals, heavy goods contents
   of wagons averaging 8 tons each   29 wagons
Empty wagons   29 wagons

These loads were in respect of passenger engines, which worked the branch.

By 1905 the length limit of freight trains permitted on the branch had increased to 40 vehicles:

Locomotive type	Goods and pick-up goods No. of wagons	Minerals and heavy goods No. of wagons	Empties No. of wagons
6-wheel coupled tender	40	30	30
Large tank engine	40	30	30
Small tank engine	40	30	30
Large 4-wheel coupled tender	40	30	30
Small 4-wheel coupled tender	40	30	30

These totals included one goods brake van.

In LNER days the length limit of freight trains on the branch were:

		No. 1 engine	No. 2 engine	No. 3 engine
Holme to Ramsey	Minerals	34	37	40
	Goods	51	55	55
	Empties	55	55	55
Ramsey to Holme	Minerals	42	45	48
	Goods	61	65	65
	Empties	65	65	65

The length limit of down trains was 55 wagons and up trains 65 wagons.

Class 'D1' tender and 'C12' tank locomotives were restricted to haul eight mineral wagons or equivalent less than a class '1' locomotive. Class '2' engines were 'J1', 'J2', 'J3' and 'J4' whilst classes 'J5', 'J6', 'J11' and 'J15' were class '3' engines.

In 1963 the running time permitted for freight trains between Holme and Ramsey North was 25 minutes. For class '7' and '8' trains a stopping allowance of three minutes per station and a starting allowance of two minutes per station was added. Because of the sharp 13 chain radius curve and rising gradient on the approach to Holme station all up goods trains over 30 wagons in length were split, the engine taking the first 30 wagons into the platform road before shunting them into the up reception sidings. The engine then returned for the remainder of the train.

Chapter Eight

Locomotives and Rolling Stock

The peaty nature of the subsoil in the fenland area served by the Ramsey North branch meant that the light formation of the permanent way precluded the use of the heaviest classes of GNR locomotives. The rural area served and relatively small loads conveyed were capable of being handled by the lighter classes of tank and tender locomotives, which with their low axle loading were ideal for the route. The GNR was fortunate in having locomotives with such universal route availability and these classes handled all passenger and freight services on the branch.

Initially the LNER classified the Ramsey North branch as route availability '4' (RA 4) with classes 'D1', 'J2', 'J6', 'J11' and 'C12' of higher route availability '5' permitted. Double-heading by all classes was prohibited. British Railways continued to classify the branch RA4 and permitted the same classes of higher route availability '5'. After the cessation of steam working the following locomotives were allowed between Holme and Ramsey North: class '3/1', later class '08' diesel shunting locomotive and classes '8/4' and '8/5' 800 hp Bo-Bo diesel electric locomotives, the latter later class '15'.

Colonel Yolland, the BoT inspector, during his examination of the railway in 1863, found that no turntable had been provided at either the junction or the terminal, and on questioning the GNR authorities learned that the company would work the line with tank locomotives which obviated the use of turntables, a decision which satisfied the inspector. The GNR initially had very few such engines and 'Little Sharpie' 2-2-2 tank locomotives worked the branch from the outset. The engines had been converted in 1852 from 2-2-2 tender locomotives built by Sharp Brothers of Manchester between 1847 and 1849, and numbered 1 to 50 inclusive. The following 14 were converted to tank locomotives between January and July 1852: Nos. 1, 2, 6, 9, 10, 18, 19, 35, 39, 40, 42, 45, 46 and 50 with a further 17, Nos. 4, 5, 7, 11, 12, 13, 15, 20, 21, 22, 28, 29, 31, 32, 33, 37 and 43, converted by the end of the year. From the date of the opening of the Ramsey line they were the only tanks engines in use on the GNR until 1865, except for three tank engines absorbed from the Nottingham to Grantham line in 1855. Unfortunately the 'Little Sharpies' had limited coal and water capacity and the necessity to replenish the Ramsey branch engine at frequent intervals rendered the class a liability and operating delays were frequent. The GNR locomotive superintendent quickly decided a change of motive power was necessary, especially with train lengths and traffic loads increasing. Thus, in defiance of the promise made to the BoT inspector, the 'Little Sharpie' tanks were withdrawn from the line and larger and more powerful tender locomotives substituted. The leading dimensions of the 'Little Sharpies' were:

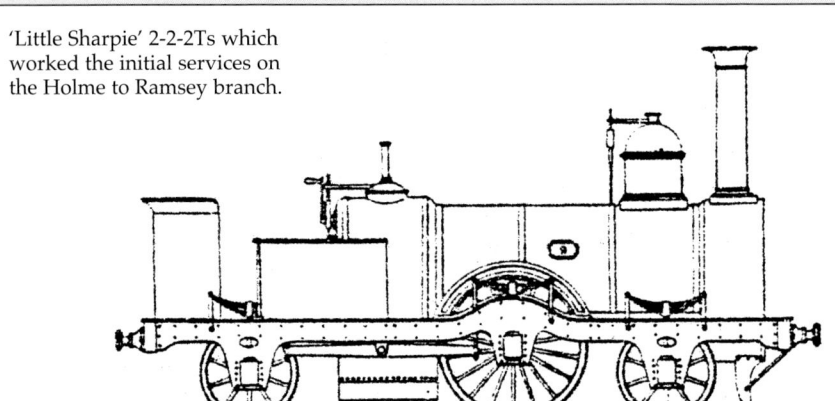

'Little Sharpie' 2-2-2Ts which worked the initial services on the Holme to Ramsey branch.

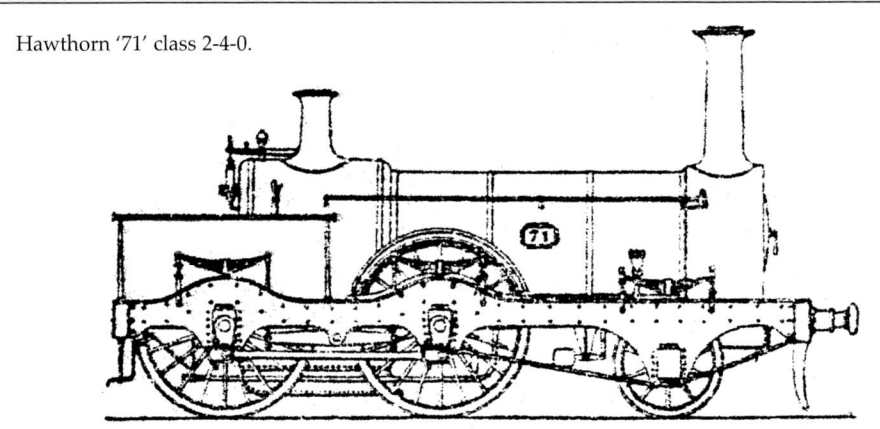

Hawthorn '71' class 2-4-0.

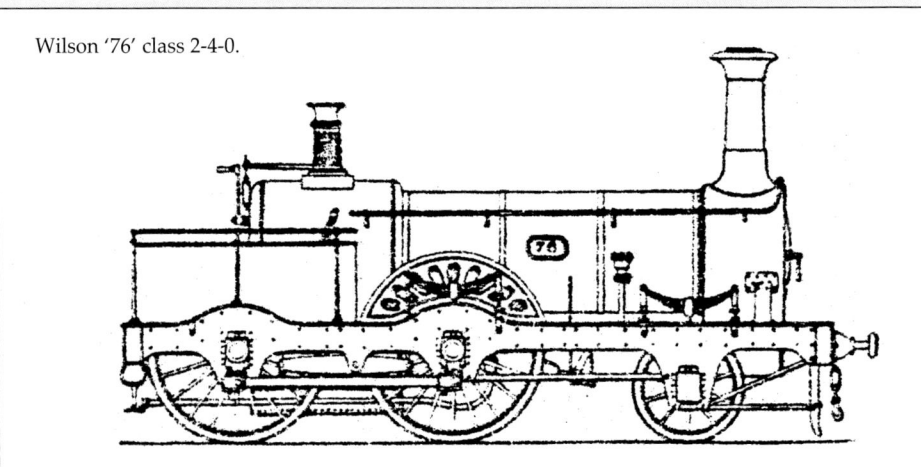

Wilson '76' class 2-4-0.

Cylinders		15 in. x 20 in.
Motion		Stephenson with slide valves
Boiler	Max. dia.	3 ft 8¾ in.
	Barrel length	10 ft 0 in.
	Firebox	3 ft 0 in.
Heating surface	Tubes 147 x 1¾ in.	690.3 sq. ft
	Firebox	57.9 sq. ft
	Total	748.2 sq. ft
Grate area		10.5 sq. ft
Leading wheels		3 ft 6 in.
Driving wheels		5 ft 6 in.
Trailing wheels		3 ft 6 in.
Wheelbase		16 ft 3 in.
Tractive effort		5,216 lb.
Axle loading	Leading wheels	8 tons 3 cwt*
	Driving wheels	7 tons 15 cwt*
	Trailing wheels	7 tons 0 cwt*
	Total	22 tons 18 cwt
Water capacity		420 gallons

* From records loadings appear to have varied between each locomotive.

From 1859 fourteen of the 'Little Sharpie' engines Nos. 3, 14, 16, 23, 24, 25, 26, 27, 30, 34, 36, 37, 44, 47 and 48 were rebuilt into 0-4-2 tender locomotives. Although many were on main line duties or working the new Sleaford line, some allocated to New England were used on the branch. The single locomotive outbased at Ramsey shed worked all the passenger and goods services and the class were in fact shown in the Appendix to the working timetables for the early 1870s as still working the line. All were condemned by 1875.

A motley collection of four-coupled tender engines were then allocated to the branch, with one locomotive outbased at Ramsey shed. The initial locomotives are believed to have included 2-4-0s built by R. & W. Hawthorn numbered in the series 71 to 75 inclusive and E. & B. Wilson & Co.'s numbered 76 to 90 inclusive. Dating from 1851 they had similar leading dimensions:

Cylinders	2 inside	16 in. x 22 in.
Motion		Stephenson with slide valves
Boiler	Max. diameter	3 ft 11 in.
	Barrel length	10 ft 0 in.
	Firebox	5 ft 2½ in.
Heating surface	Tubes 157 x 2 in.	802.0 sq. ft
	Firebox	102.0 sq. ft
	Total	904.0 sq. ft
Boiler pressure		120 psi
Leading wheels		4 ft 0 in.
Coupled wheels		6 ft 0 in.
Wheelbase	Locomotive	15 ft 0 in.
Weight in working order		
	Locomotive	27 tons 18 cwt
	Water capacity	1,300 gallons
	Coal capacity	5 tons

At this period a few of the 'Small Hawthorn' 2-2-2 tender locomotives, built by R. & W. Hawthorn & Co. between October 1848 and October 1850, and numbered in the series 51 to 70 inclusive, had been displaced from main line working, and were rebuilt for further use on lighter secondary expresses and minor branch lines. At least two worked on the Ramsey branch in their final years when allocated to the Peterborough district until withdrawal in 1872. The leading dimensions were:

Cylinders	2 inside	16 in. x 21 in.
Motion		Stephenson with slide valves
Boiler	Max. dia. outside	4 ft 0 ½ in.
	Barrel length	10 ft 0 in.
	Firebox length outside	4 ft 9 in.
Heating surface	Tubes 173 x 1¾ in.	839.0 sq. ft
	Firebox	68.0 sq. ft
	Total	907.0 sq. ft
	Grate area	15.0 sq. ft
Boiler pressure		100 psi
Leading wheels		3 ft 10 in.
Driving wheels		6 ft 1 in.
Trailing wheels		3 ft 10 in.
Engine wheelbase		13 ft 9 in.
Weight empty		28 tons 1 cwt*
Water capacity		1,200 gallons
Coal capacity		2 tons

* Locomotive only.

Between April 1848 and June 1849, Hawthorn had supplied to the GNR 15 outside frame 0-4-2 'luggage' engines, Nos. 101 to 115 inclusive. From the mid-1860s most had been displaced from front line goods work by 0-6-0 locomotives and were found lighter work on branch lines. Fourteen of the class, were placed on the duplicate list between 1881 and 1885, whilst withdrawals began in 1887. Nos. 110A and 113A were allocated to Peterborough in their last years before scrapping in 1896 and spent some time on Ramsey branch services. All were withdrawn by 1899. The principal dimensions were:

Cylinders	2 inside	16 in. x 24 in.
Motion		Stephenson with slide valves
Boiler	Max. dia. outside	3 ft 11 in.
	Barrel length	10 ft 0 in.
	Firebox outside length	4 ft 9 in.
	Heating surface	
	Tubes 148 x 1⅞ in.	744.5 sq. ft
	Firebox	74.0 sq. ft
	Total	818.5 sq. ft
Grate area		10.0 sq. ft
Boiler pressure		120 psi
Coupled wheels		5 ft 0 in.
Trailing wheels		3 ft 6 in.
Engine wheelbase		14 ft 0 in.
Weight in working order		28 tons 0 cwt*

* Locomotive only.

LOCOMOTIVES AND ROLLING STOCK 125

The next class to work on the Ramsey branch, were some of Patrick Stirling's 2-4-0 tender locomotives, with 6 ft 6 in. driving wheels. Stirling assumed control of the GNR locomotive department in 1866 and during the following two years introduced into traffic twenty 2-4-0 passenger tender locomotives numbered 280 to 299 inclusive, the first 10 built by the Avonside Engine Co. and the remainder by the Yorkshire Engine Co. Another two were built at Doncaster Works in 1871. Between 1874 and 1879 fourteen of the '86' series locomotives entered service from Doncaster to be followed by eight of the '223' series in 1880/81. The '78' series followed in 1882/83, with six built at Doncaster and six by Kitson & Co. Kitson supplied nine further locomotives in 1884 under the '707' series, with Doncaster building the balancing two of the order. Between 1884 and 1895 Doncaster Works turned out another 72 locomotives of the '206' series, whilst the balance of the 149 2-4-0s were produced as the '1061' series by Ivatt in 1897. The various classes only worked the Huntingdonshire branch after being displaced from main line duties and never gained a monopoly on the Ramsey line. From about 1876 some of the Peterborough district engines were rostered to work the branch goods and passenger services, turn and turn about with Stirling's '18' and '103' class 0-4-2 locomotives but were never popular. The locomotives were later designated as GNR classes 'E1', 'E2' and 'E3', and those allocated to the Peterborough district included:

Class	Later GNR class	No.	Built	Withdrawn
86	E2	84	September 1874	October 1909
86	E1	89	July 1874	June 1913
86	E1	90	November 1874	November 1922
86	E1	99	October 1879	February 1913
206	E1	210	October 1888	November 1922
206	E2	211	February 1886	March 1912
206	E2	213	August 1889	September 1912
206	E2	214	September 1889	October 1908
206	E1	217	March 1886	April 1921
206	E1	225	September 1886	January 1922
223	E1	226	March 1881	February 1912
280	E3	262A	November 1871	February 1909
280	E3	294A	August 1868	April 1912
206	E1	753	February 1885	January 1923
206	E1	754	February 1885	June 1921
206	E1	755	November 1886	February 1923
206	E2	814	May 1889	November 1927
206	E2	815	July 1889	April 1912
206	E2	817	October 1889	May 1912
206	E1	819	December 1889	July 1922
206	E2	820	December 1889	November 1914
206	E1	884	December 1892	September 1925
206	E2	885	December 1892	January 1923
206	E2	886	April 1893	December 1913
206	E2	890	October 1893	September 1912
206	E1	893	December 1893	January 1923
206	E1	894	December 1893	July 1921

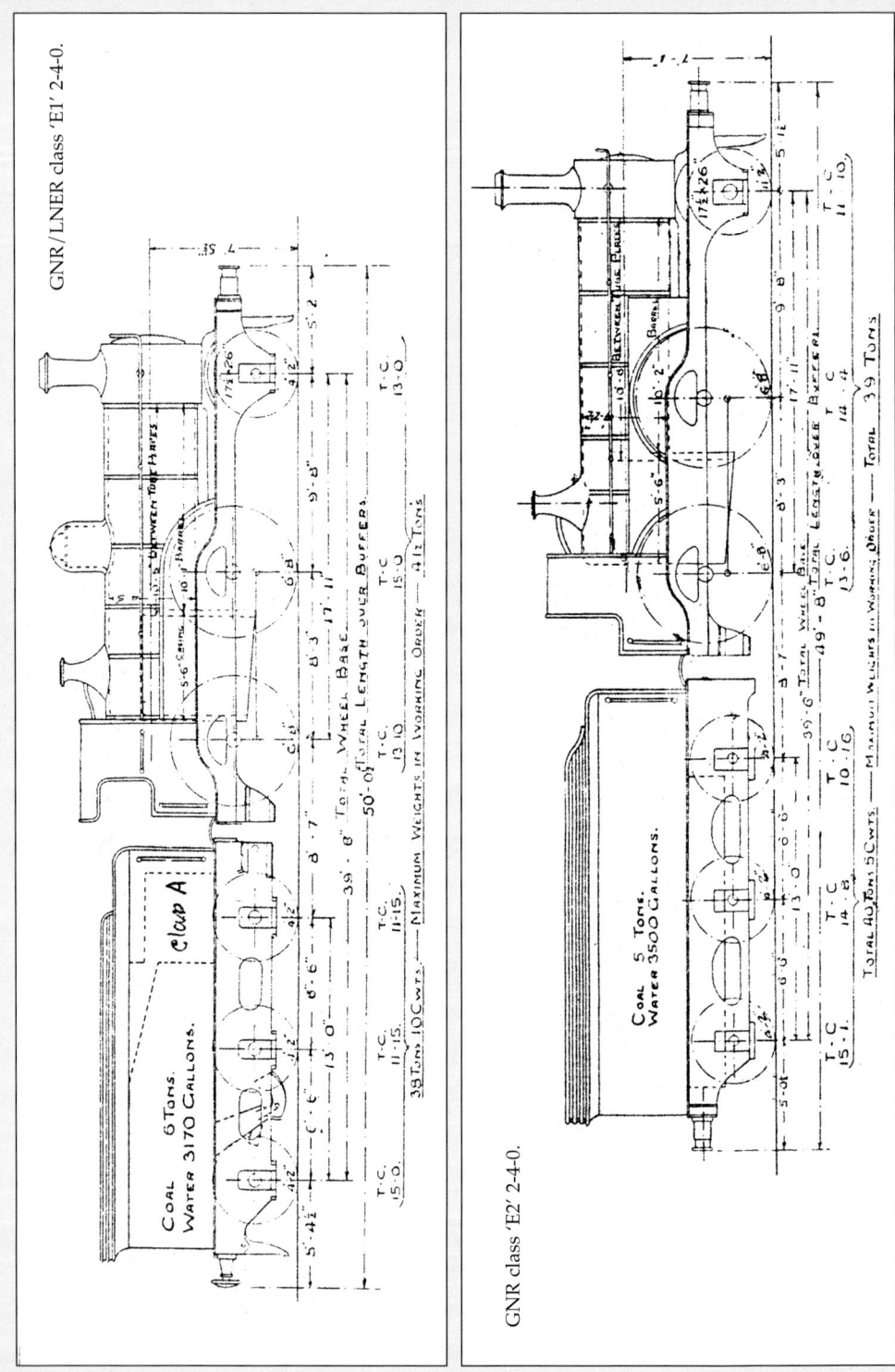

LOCOMOTIVES AND ROLLING STOCK

Class	Later GNR class	No.	Built	Withdrawn
206	E2	899	April 1894	December 1921
206	E2	900	April 1894	November 1912
206	E2	994	September 1894	August 1924
206	E1	995	November 1894	June 1923
206	E1	996	December 1894	July 1922
206	E2	997	May 1895	July 1912
206	E1	998	May 1895	January 1924
1061	E1	1068	April 1897	July 1925
1061	E1	1069	April 1897	August 1924

The leading dimensions of the '280' series were:

Cylinders	2 inside	17 in. x 24 in. (later 17½ in. x 24 in.)
Motion		Stephenson with slide valves
Boiler	Max. diameter	4 ft 0½ in.
	Barrel length	10 ft 2 in.
	Firebox	5 ft 6 in.
Heating surface	Tubes 206 x 1¾ in.	991.5 sq. ft
	Firebox	94.0 sq. ft
	Total	1,085.5 sq. ft
Grate area		16.25 sq. ft
Boiler pressure		130 psi
Leading wheels		4 ft 1 in.
Coupled wheels		6 ft 7 in.
Tractive effort		9,826 lb.
Length over buffers		49 ft 6 in.*
Wheelbase	Locomotive	17 ft 9 in.
	Tender	13 ft 0 in.
	Total	39 ft 4 in.
Weight in working order	Locomotive	38 tons 0 cwt
	Tender	34 tons 18 cwt
Max. axle loading		14 tons 0 cwt
Water capacity		2,500 gallons
Coal capacity		6 tons

* Engine and tender.

The '86' series had the following detailed differences:

Cylinders	2 inside	17½ in. x 26 in.
Heating surface	Tubes 169 x 1⅝ in.	743.0 sq. ft
	Firebox	100.0 sq. ft
	Total	843.0 sq. ft
Boiler pressure		140 psi
Tractive effort		12,147 lb.
Length over buffers		49 ft 8 in.*
Weight		38 tons 16 cwt
Max. axle loading		14 tons 0 cwt
Water capacity		2,800 gallons
Coal capacity		4 tons

* Engine and tender.

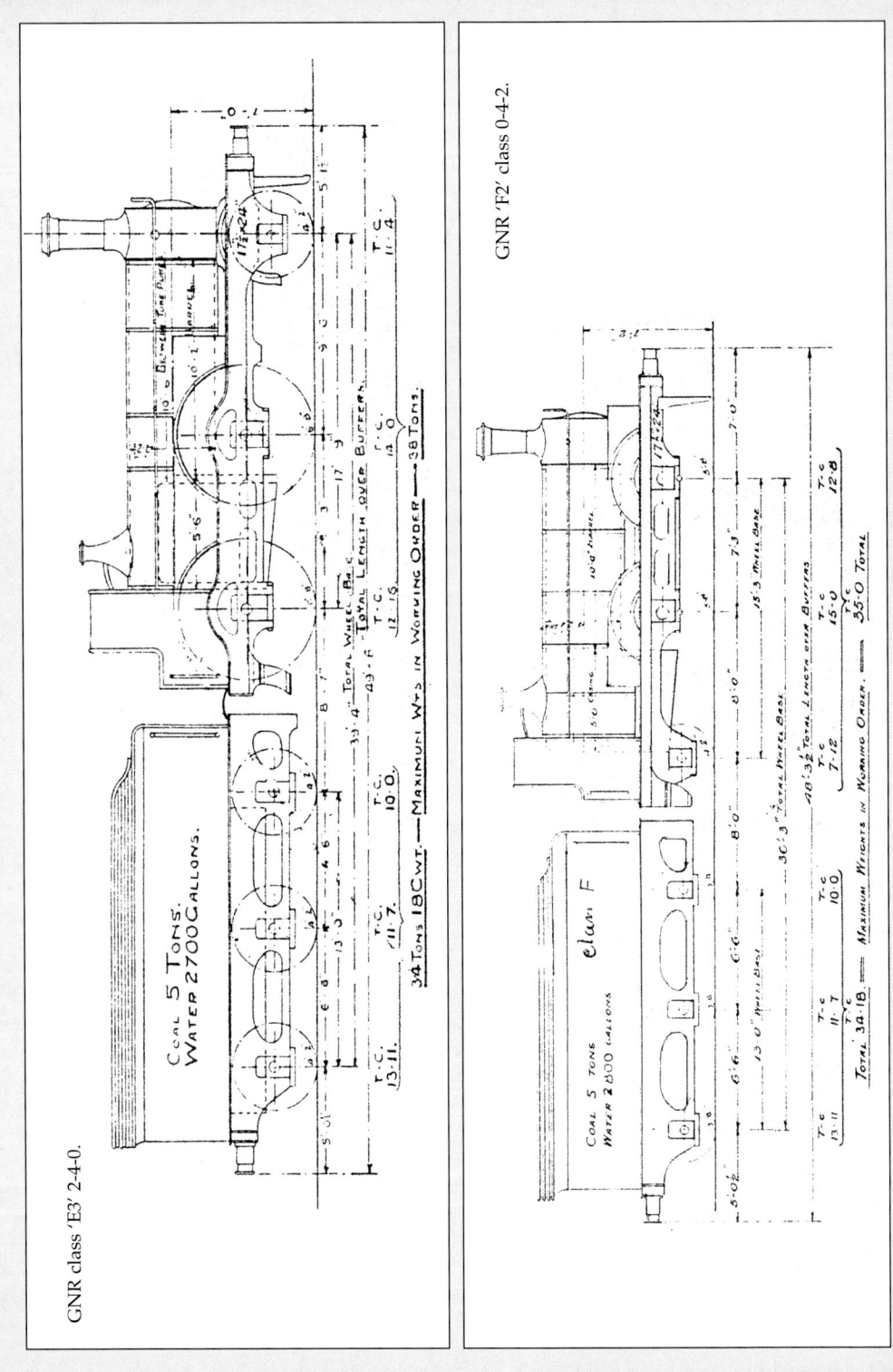

Yet further differences came with the building from No. 757 on:

Heating surface	Tubes 174 x 1¾ in.	836.0 sq. ft
	Firebox	92.0 sq. ft
	Total	928.0 sq. ft
Boiler pressure		160 psi
Leading wheels		4 ft 1 ½ in.
Coupled wheels		6 ft 7 ½ in.
Tractive effort		13,883 lb.
Length over buffers		50 ft 0½ in*
Weight		39 tons 0 cwt
Max. axle loading		14 tons 4 cwt
Water capacity		3,500 gallons
Coal		5 tons

* Engine and tender

The dimensions of the final development '1061' series were:

Cylinders	2 inside	17 ½ in. x 26 in.
Motion		Stephenson with slide valves
Boiler	Max. diameter outside	4 ft 5 in.
	Barrel length	10 ft 1 in.
	Firebox outside length	5 ft 6 in.
	Heating surface	
	Tubes 213 x 1¾ in.	1,016.0 sq. ft
	Firebox	103.0 sq. ft
	Total	1,119.0 sq. ft
Grate area		16.25 sq. ft
Boiler pressure		170 psi
Leading wheels		4 ft 2 in.
Coupled wheels		6 ft 8½ in.
Tractive effort		14,751 lb.
Length over buffers		50 ft 0 ½ in.*
Weight		41 tons 10 cwt
Max. axle loading		15 tons 0 cwt
Water capacity		3,170 gallons
Coal		6 tons

* Engine and tender.

In 1868 the first of Patrick Stirling's 0-4-2 locomotives designed specially for mixed traffic work, entered service as GNR class '18'. Such was the success of the design that a total of 154 were built between 1868 and 1895, including the later development '218' and '103' series. Most were built at Doncaster but Nos. 551 to 580 were constructed by Sharp, Stewart, and Nos. 581 to 600 by Kitson & Co. Over the years several detailed differences and improvements were introduced and Ivatt rebuilt 10 engines between 1900 and 1902. Whilst they operated all over the GNR system, several were allocated to Peterborough and these worked across the Ramsey branch as part of their duties. In the later GNR locomotive classification, the '218' series locomotives became 'F1' class, whilst the remaining engines were designated class 'F2' with the Ivatt rebuilds as 'F3'.

GNR 'F2' class 0-4-2s were often utilized on the Ramsey branch services. Here a member of the class, No. 205, is working a local train at Basford, on the Derby to Nottingham line in February 1910. The locomotive was originally one of the class '18' series. *Author's Collection*

The following 'F2' and 'F3' class locomotives are known to have handled the Ramsey services:

Class	No.	Built	Withdrawn
18	19	June 1869	October 1907
18	27	December 1869	May 1908
18	58A	May 1870	December 1919
18	75	September 1872	May 1910
103	110	October 1885	March 1921
103	112	December 1883	November 1918
18	200	November 1870	June 1921
103	326	June 1888	March 1921
18	554	January 1876	April 1908
18	557	March 1876	August 1912
18	589	May 1876	June 1919
18	592A	June 1876	April 1921

Of the above locomotives, Ivatt rebuilt the following with 4 ft 5 in. diameter domed boilers:

58A rebuilt October 1914
110 rebuilt October 1914
200 rebuilt December 1915
589 rebuilt October 1914
592A rebuilt November 1914

The principal dimensions of the class '18' 0-4-2 locomotives as rebuilt with a 4 ft 2½ in. diameter boiler were:

Cylinders	2 inside	17½ in. x 24 in.
Motion		Stephenson with slide valves
Boiler	Max. diameter outside	4 ft 2½ in.
	Barrel length	10 ft 0 in.
	Firebox	5 ft 6in.

Heating surface	Tubes 169 x 1⅝ in.	743.0 sq. ft
	Firebox	94.5 sq. ft
	Total	837.5 sq. ft
Firebox grate area		16.0 sq. ft
Boiler pressure		140 psi
Coupled wheels		5 ft 7½ in.
Trailing wheels		3 ft 7½ in.
Tractive effort		12,957 lb.
Length		23 ft 0 in.
Length over buffers		48 ft 3½ in.*
Wheelbase	Locomotive	15 ft 2 in.
Weight in working order	Locomotive	32 tons 3 cwt
Water capacity		2,400 gallons
		2,600 gallons
Coal capacity		5 tons

* Engine and tender

The '103' series locomotives had the following detailed differences when built:

Heating surface	Tubes 186 x 1¾ in.	879.0 sq. ft
	Firebox	94.0 sq. ft
	Total	973.0 sq. ft
Boiler pressure		160 psi
Trailing wheels		4 ft 1½ in.
Tractive effort		14,809 lb.
Length		24 ft 0 in.
Engine wheelbase		15 ft 3 in.
Weight in working order		35 tons 2 cwt
Water capacity		2,500 gallons
		2,800 gallons
Coal capacity		6 tons

The engines rebuilt by Ivatt had the following leading dimensions:

Cylinders	2 inside	17½ in. x 24 in.
Motion		Stephenson with slide valves
Boiler	Max. outside diameter	4 ft 5 in.
	Barrel length	10 ft 1 in.
	Firebox outside length	5 ft 6 in.
Heating surface	Tubes 215 x 1¾ in.	1,020.7 sq. ft
	Firebox	103.1 sq. ft
	Total	1,123.8 sq. ft
Grate area		17.8 sq. ft
Boiler pressure		160 psi
Coupled wheels		5 ft 7½ in.
Trailing wheels		3 ft 7½ in.
		4 ft 1 ½ in.
Tractive effort		14,809 lb.
Water capacity		2,600 gallons
Coal capacity		5 tons

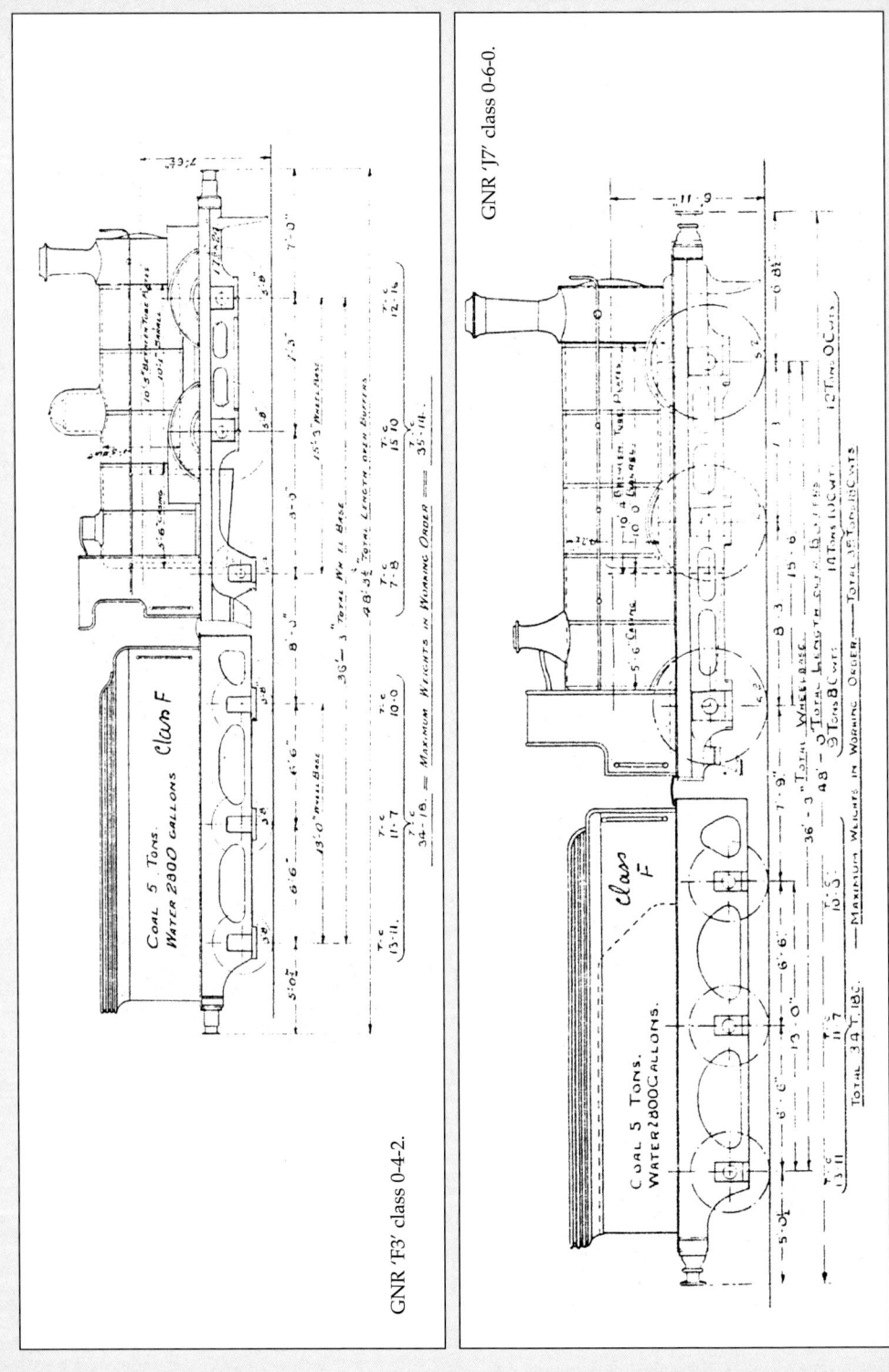

LOCOMOTIVES AND ROLLING STOCK

In 1867 Patrick Stirling introduced into service his first 0-6-0 locomotive design on the GNR. Initially five were built by John Fowler & Co. and 10 more by Neilson & Co., with a further five from Fowler the following year. Known as the '474' series the 20 engines were numbered 474 to 493 inclusive. Between 1869 and 1873 another 15 in the '369' series, built at Doncaster Works entered traffic. Under the GNR engine classification scheme the engines were classified 'J7' in 1900 and the locomotives allocated to New England shed, once displaced from main line duties, worked freight and engineers trains on the Ramsey branch. These included:

Series	No.	Built	Withdrawn
369	151	April 1873	January 1908
369	186	November 1873	April 1914
369	311	April 1872	November 1918
369	377	November 1869	December 1910
369	380	March 1870	November 1912
474	478	January 1868	November 1912
474	490	November 1867	December 1909

The principal dimensions of the 'J7' class were:

Cylinders	2 inside	17 in. x 24 in.
Motion		Stephenson with slide valves
Boiler	Max. diameter	4 ft 0½ in.
	Barrel length	10 ft 0 in.
	Firebox	5 ft 6 in.
Heating surface	Tubes 206 x 1¾ in.	985.50 sq. ft
	Firebox	94.25 sq. ft
	Total	1,079.75 sq. ft
Firebox grate area		16.25 sq. ft
Boiler pressure		150 psi
Coupled wheels		5 ft 1 in.
Tractive effort		14,440 lb.
Length over buffers		48 ft 0 in.*
Wheelbase	Locomotive	15 ft 6 in.
	Tender	13 ft 0 in.
	Total	36 ft 3 in.
Weight in working order	Locomotive	32 tons 11 cwt
	Tender	34 tons 18 cwt
Max. axle load		14 tons 10 cwt
Water capacity		2,000 gallons
Coal capacity		4 tons 15 cwt

* Engine and tender.

Three classes to be associated with the Ramsey line for many years were Stirling's and Ivatt's GNR class 'J4', 'J5' and 'J6' 0-6-0s. Initially developed in 1873, the 38 of the '171' series were all built at Doncaster Works from that date until 1881. The following '716' series, comprised 10 locomotives built by Vulcan Foundry between August and November 1882, and 20 built simultaneously by Dübs & Co. from June to November 1882. The balancing 10 of this series were built at Doncaster Works in 1886. The 62 members of the '322' series were all built at Doncaster between 1887 and 1894, whilst the 15 '1031' series were supplied by Dübs & Co. in 1896. The locomotives were later designated GNR class 'J6'. When Ivatt took up his appointment as locomotive superintendent in March 1896, he adopted the 10 '1081' series built in the same year at Doncaster, for further development, and between 1897 and 1901 arranged for another 133 engines to be constructed. Starting with 10 locomotives in the '1091' series built by Dübs in 1897/98, the '315' series followed with 10 built at Doncaster in 1898 and 15 by Dübs in 1898/99. The '343' series completed the order, with 40 built at Doncaster in 1899 and 1900, 25 from Kitson & Co. in 1900 and 13 from Dübs in 1901. These 143 locomotives were designated class 'J5' in 1900 and over the years 125 of the earlier 'J6s' were rebuilt to conform to class 'J5'.

In May 1912 Gresley commenced equipping some of the class with 4 ft 8 in. diameter boilers and 71 locomotives were converted up to Grouping, with the LNER converting a further 82 engines. These rebuilds became GNR class 'J4'. The 228 survivors of the 'Standard Goods' class were reclassified by the LNER, the GNR 'J4' becoming LNER class 'J3' and GNR class 'J5', LNER 'J4'. The various locomotives allocated to Peterborough over the years had regularly worked the Ramsey passenger and goods services, especially after the closure of Ramsey shed, and '171' series No. 340 was involved in the collision at Ramsey on 15th November, 1893. From the turn of the century 16 of the two classes were allocated to the Peterborough district, and included in their duties were the daily goods workings from New England Yard to Ramsey. Between the two world wars New England often placed one of the class on excursion trains from Ramsey North, St Mary's and Holme to Skegness and the RA3 route availability was well within the limits imposed by the civil engineer on the branch. Although 16 of the combined classes were extant at New England at nationalization, the class quickly succumbed to the cutters torch and the last 'J4' worked to Ramsey North in 1951. In her final days before withdrawal in November 1942, No. 4154 was the regular engine on the branch trains. The undermentioned are known to have worked the branch services.

LNER class	GNR No.	LNER 1924 No.	LNER 1946 No.	BR No.	Withdrawn
J3	390	3390	–	–	July 1936
J3	1082	4082	–	–	April 1935
J3	1104	4104	–	–	October 1935
J3	1105	4105	4119	64119	July 1951
J3	1108	4108	–	–	September 1935
J4	181	3181	–	–	August 1926
J4	193	–	–	–	February 1927
J4	198	3198	–	–	May 1927
J4	312	–	–	–	April 1926
J4	339	3339	–	–	June 1928
J4	340	–	–	–	May 1919
J4	385	3385	–	–	January 1929
J4	392	3392	–	–	February 1932
J4	727	–	–	–	August 1926
J4	1087	4087	–	–	October 1931

LNER class	GNR No.	LNER 1924 No.	LNER 1946 No.	BR No.	Withdrawn
J4	1102	4102	–	–	November 1932
J3	306	3306	4140	64140	December 1954
J3	717	3717	–	–	November 1939
J3	1103	4103	4118	64118	June 1952
J3	1109	4109	4122	64122	May 1953
J3	1163	4163	–	–	May 1938
J4	332	3332	4133	64133	January 1953
J4	1011	4011	–	–	January 1940
J4	1037	4037	–	–	February 1939
J4	1090	4090	4112	64112	December 1951
J4	1154	4154	–	–	November 1942
J4	1160	4160	4151	–	March 1951

The standard dimensions of the LNER 'J3' class were:

Cylinders	2 inside	17½ in. x 26 in.
Motion		Stephenson with slide valves
Boiler	Max. diameter	4 ft 8 in.
	Barrel length	10 ft 1 in.
	Firebox	5 ft 6 in.
Heating surface	Tubes 238 x 1¾ in.	1,130.0 sq. ft
	Firebox	105.0 sq. ft
	Total	1,235.0 sq. ft
Firebox grate area		16.25 sq. ft
Boiler pressure		175 psi
Coupled wheels		5 ft 2 in.
Tender wheels		4 ft 2 in.
Length over buffers		49 ft 6 in.*
Tractive effort		19,105 lb.
Wheelbase	Locomotive	15 ft 6 in.
	Tender	13 ft 0 in.
	Total	36 ft 11½ in.
Weight in working order	Locomotive	42 tons 12 cwt
	Tender	38 tons 10 cwt
	Total	81 tons 2 cwt
Max. axle loading		16 tons 0 cwt
Water capacity		3,170 gallons
Coal capacity		6 tons 0 cwt

* Engine and tender

The principal dimensions of the LNER 'J4' class were:

Cylinders	2 inside	17 ½ in. x 26 in.
Motion		Stephenson with slide valves
Boiler	Max. diameter	4 ft 5 in.
	Barrel length	10 ft 1 in.
	Firebox	5 ft 6 in.
Heating surface	Tubes 213 x 1¾ in.	1,016.0 sq. ft
	Firebox	103.0 sq. ft
	Total	1,119.0 sq. ft

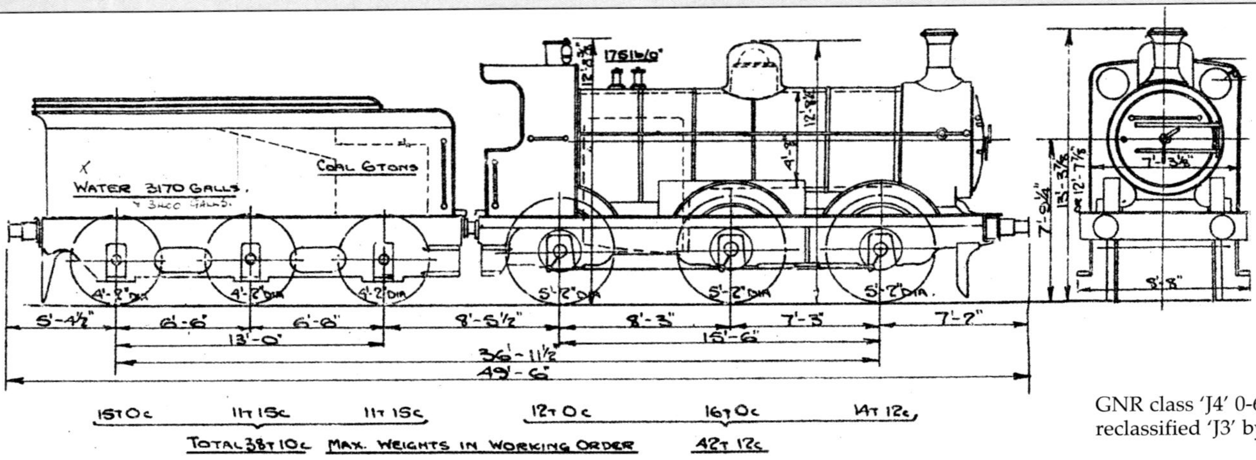

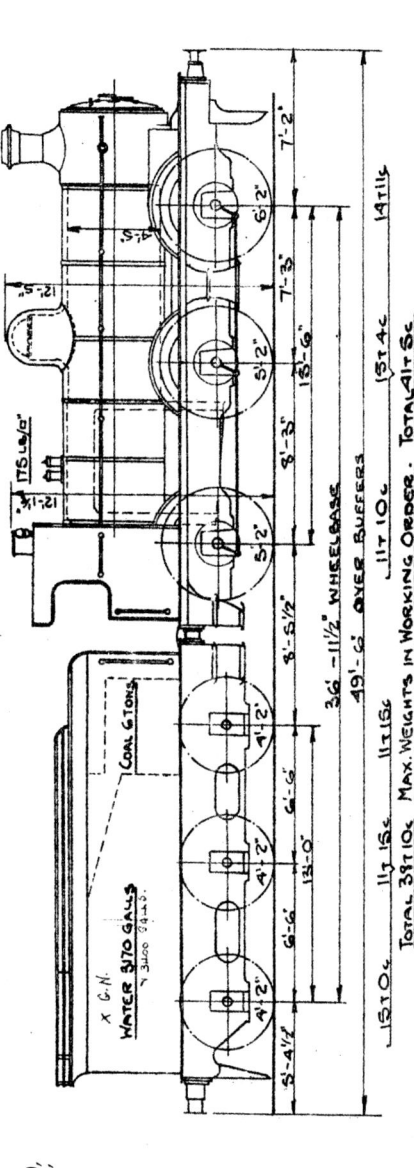

GNR class 'J4' 0-6-0, later reclassified 'J3' by the LNER.

GNR class 'J5' 0-6-0, later reclassified 'J4' by the LNER.

Firebox grate area		16.25 sq. ft
Boiler pressure		170 psi
Coupled wheels		5 ft 2 in.
Tender wheels		4 ft 2 in.
Tractive effort		18,560 lb.
Length over buffers		49 ft 2 in.*
Wheelbase	Locomotive	15 ft 6 in.
	Tender	13 ft 0 in.
	Total	36 ft 11½ in.
Weight in working order	Locomotive	41 tons 5 cwt
	Tender	34 tons 18 cwt
	Total	76 tons 3 cwt
Max. axle loading		15 tons 4 cwt
Water capacity		2,800 gallons
Coal capacity		5 tons 0 cwt

* Engine and tender

A class, which saw occasional use on the Ramsey branch, was Ivatt's 'J21' class 0-6-0s later classified 'J1' by the LNER. Originally built in 1908 for working fast freight trains, the 15 members of the class were soon deposed from such duties in 1912 and 1913 by Gresley's more powerful 2-6-0s, later LNER class 'K2'. Thereafter the 'J21s' were transferred to the less exacting work on cross-country, secondary and branch freight services as well as a number of passenger train diagrams. Whilst at New England in the years after World War I, No. 13 worked occasionally between Holme and Ramsey and in 1930 No. 3015 was recorded working the branch services on a regular basis, deputizing for the usual 'C12' class 4-4-2T The 'J1' class thereafter had little connection with the branch, as the LNER 'J4' or 'J6' class 0-6-0s hauled most of the services. However, soon after World War II four of the class, by now renumbered 5002/4/5/6 were based at New England and although much of their work involved duties on the Midland & Great Northern line Nos. 5002 and 5005 were recorded working the daily services to Ramsey North on various occasions between 1947 and 1951. Of the 'J1' class locomotives known to have worked on the branch, disposals were as follows:

GNR No.	LNER 1924 No.	LNER 1946 No.	BR No.	Withdrawn
3	3003	5002	65002	August 1954
6	3006	5005	65005	May 1952
13	3013	5012	–	August 1947
15	3015	5014	65014	August 1953

The leading dimensions of the LNER 'J1' class were:

Cylinders	2 inside	18 in. x 26 in.
Motion		Stephenson with slide valves
Boiler	Max. diameter	4 ft 8 in.
	Barrel length	10 ft 1 in.
	Firebox	6 ft 4 in.

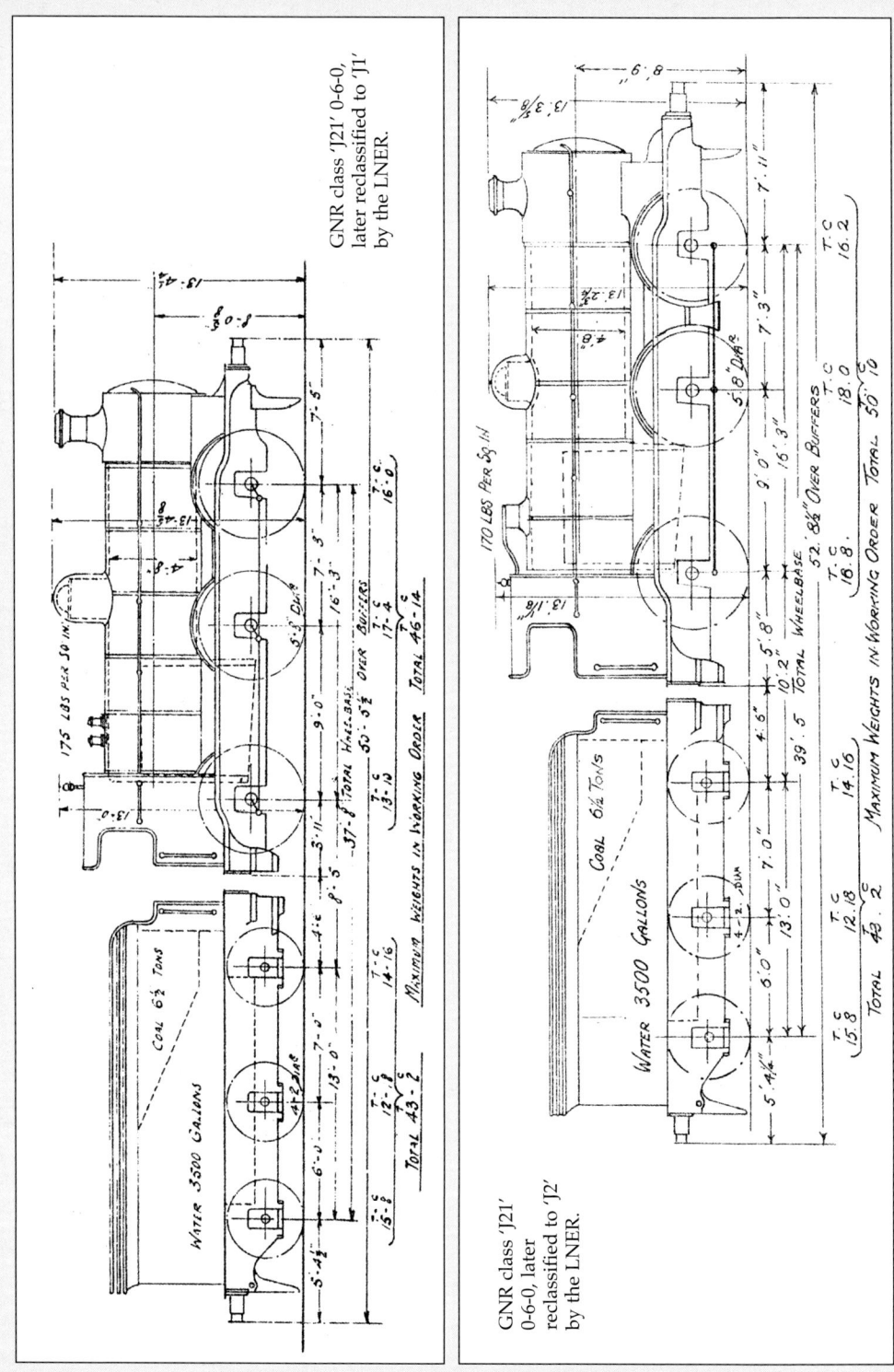

Heating surface	Tubes 238 x 1¾ in.	1,130.0 sq. ft
	Firebox	120.0 sq. ft
	Total	1,250.0 sq. ft
Firebox grate area		19.0 sq. ft
Boiler pressure		175 psi
Coupled wheels		5 ft 8 in.
Tender wheels		4 ft 2 in.
Tractive effort		18,427 lb.
Length over buffers		50 ft 5 ½ in.*
Wheelbase	Locomotive	16 ft 3 in.
	Tender	13 ft 0 in.
	Total	37 ft 8 in.
Weight in working order	Locomotive	46 tons 14 cwt
	Tender	43 tons 2 cwt
	Total	89 tons 16 cwt
Max. axle loading		17 tons 4 cwt
Water capacity		3,500 gallons
Coal capacity		6 tons 10 cwt

* Engine and tender

Another of the GNR 0-6-0 classes to work on the Huntingdonshire branch also Ivatt's 'J21' class, but later designated class 'J2' by the LNER. Ten locomotives were introduced in 1912 and originally worked the fast braked goods trains between London and York. Within two years the class was also displaced from the express services by Gresley's 2-6-0 tender locomotives, and Nos. 72 to 78 inclusive were transferred to New England with the other three going to Doncaster. Soon after Grouping the newly numbered 3076, 3077 and 3078 were transferred to Colwick. During their stay at New England the 'J2s' were occasionally drafted to the Ramsey North branch goods trips working turn and turn about with the 'J1', 'J3', 'J4' and 'J6' classes, although their main allocated work involved local pick-up goods and passenger duties on the Lincolnshire lines.

GNR No.	LNER 1924 No.	LNER 1946 No.	BR No.	Withdrawn
72	3072	5016	65016	October 1953
73	3073	5017	65017	January 1954
74	3074	5018	65018	November 1953
75	3075	5019	65019	March 1953
76	3076	5020	65020	July 1954
77	3077	5021	–	November 1950
78	3078	5022	65022	December 1953

The principal dimensions of the 'J2' class were:

Cylinders	2 inside	19 in. x 26 in.
Motion		Stephenson with 8 in. piston valves
Boiler		
	Max. diameter	4 ft 8 in.
	Barrel length	10 ft 1 in
	Firebox	6 ft 4 in.

'J6' class 0-6-0 No. 3528 waits in the Ramsey North branch platform at Holme in 1936. The van behind the tender was provided for perishable traffic and parcels. Note the ornate station nameboard and the post oil lamp on the narrow island platform.
Author's Collection

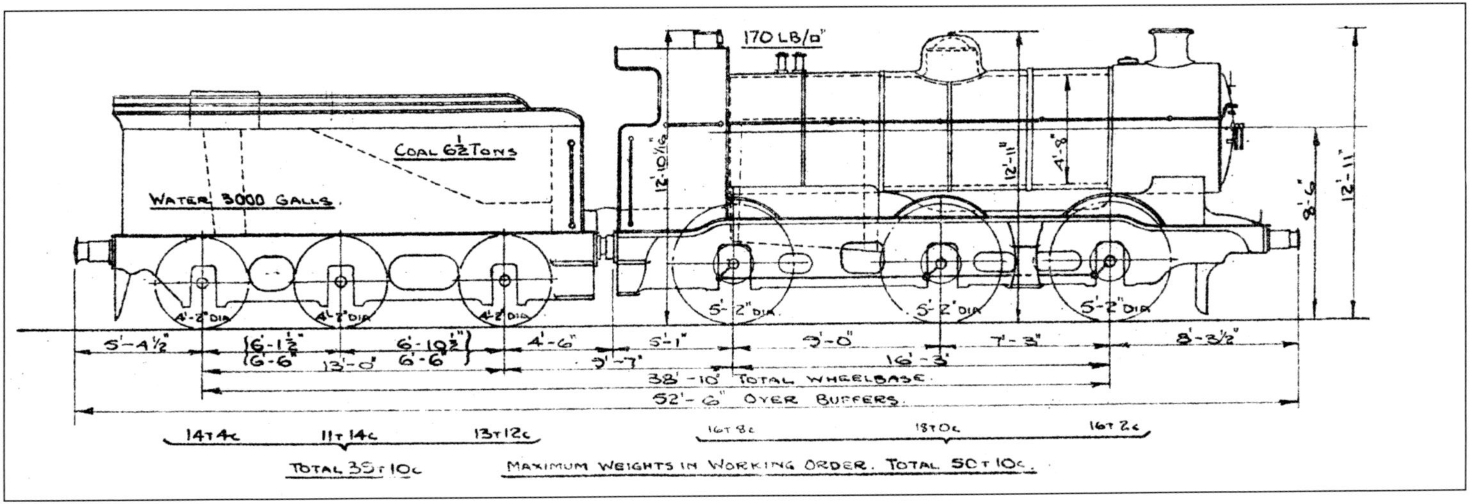

GNR class 'J22' 0-6-0, later reclassified to 'J6' by the LNER worked the Ramsey branch goods services for many years.

Heating surface	Tubes 118 x 1¾ in.	562.0 sq. ft	
	Firebox	118.0 sq. ft	
	Flues 18 x 5¼ in.	257.0 sq. ft	
	Total evaporative	937.0 sq. ft	
	Superheater 18 x 1¼ in.	192.0 sq. ft	
	Total	1,129.0 sq. ft*	
Firebox grate area		19.0 sq. ft	
Boiler pressure		170 psi	
Coupled wheels		5 ft 8 in.	
Tender wheels		4 ft 2 in.	
Tractive effort		19,945 lb.	
Length over buffers		52 ft 8 ½ in.†	
Wheelbase	Locomotive	16 ft 3 in.	
	Tender	13 ft 0 in.	
	Total	39 ft 5 in.	
Weight in working order	Locomotive	50 tons 10 cwt	
	Tender	43 tons 2 cwt	
	Total	93 tons 12 cwt	
Max. axle loading		18 tons 0 cwt	
Water capacity		3,500 gallons	
Coal capacity		6 tons 10 cwt	

* With superheater. † Engine and tender.

By far the most prolific of the GNR 0-6-0s to work the Holme to Ramsey services were the Ivatt 'J22' class, later designated 'J6' by the LNER. A total of 110 engines built between 1911 and 1922, were the final development of the Ivatt 0-6-0s with traditional Doncaster outline and were fitted with superheated boilers. New England shed always had a large allocation of the class and from their introduction to the Peterborough district until their final withdrawal from traffic the 'J6s' were firm favourites with engine crews. From February 1931 the class virtually took over the monopoly of the remaining passenger and daily freight services. The locomotive allocated to work the line hauled the morning freight from New England or Yaxley Yard to Holme before hauling the 5.58 am goods train to Ramsey North. The engine then covered the remaining passenger and freight services for the rest of the day before returning to New England shed in the evening. The 'J6s' were finally ousted from the line in 1959 when London Midland class '4' 2-6-0s, displaced by the closure of the Midland & Great Northern line, were transferred to New England to take over much of the work formerly allocated to these engines. Locomotives known to have worked the Ramsey North branch included:

GNR No.	LNER 1924 No.	LNER 1946 No.	BR No.	Withdrawn
522	3522	4171	64171	September 1961
527	3527	4176	64176	March 1959
528	3528	4177	64177	February 1962
531	3531	4180	64180	March 1960
532	3532	4181	64181	August 1959
534	3534	4183	64183	November 1958
535	3535	4184	64184	November 1959
537	3537	4186	64186	January 1958

GNR No.	LNER 1924 No.	LNER 1946 No.	BR No.	Withdrawn
538	3538	4187	64187	January 1958
540	3540	4189	64189	October 1958
542	3542	4191	64191	February 1962
543	3543	4192	64192	April 1960
548	3548	4197	64197	October 1959
554	3554	4203	64203	June 1962
555	3555	4204	64204	December 1957
558	3558	4207	64207	October 1959
562	3562	4211	64211	March 1958
567	3567	4216	64216	December 1958
568	3568	4217	64217	January 1959
571	3571	4220	64220	June 1958
576	3576	4225	64225	July 1958
578	3578	4227	64227	July 1958
579	3579	4228	64228	August 1959
586	3586	4235	64235	December 1959
589	3589	4238	64238	October 1959
593	3593	4242	64242	August 1955
594	3594	4243	64243	June 1958
596	3596	4245	64245	February 1962
597	3597	4246	64246	June 1959
600	3600	4249	64249	October 1958
603	3603	4252	64252	July 1958
605	3605	4254	64254	October 1959
608	3608	4257	64257	June 1960
621	3621	4260	64260	March 1961
622	3622	4261	64261	March 1959
623	3623	4262	64262	February 1959
624	3624	4263	64263	July 1958
625	3625	4264	64264	January 1958
626	3626	4265	64265	July 1961
627	3627	4266	64266	April 1959
634	3634	4273	64273	December 1959
636	3636	4275	64275	August 1958
639	3639	4278	64278	March 1961

For many years Nos. 3528 and 3562 spent most of their time on the Ramsey North branch duties. The leading dimensions of the class were:

Cylinders	2 inside	19 in. x 26 in.
Motion		Stephenson with 8 in. piston valves
Boiler	Max. diameter	4 ft 8 in.
	Barrel length	10 ft 1 in.
	Firebox	6 ft 4 in.
Heating surface	Tubes 118 x 1¾ in.	562.0 sq. ft
	Flues 18 x 5¼ in.	257.0 sq. ft
	Firebox	118.0 sq. ft
	Total evaporative	937.0 sq. ft
	Superheater 18 x 1¼ in.	192.0 sq. ft
	Total	1,129.0 sq. ft*

Firebox grate area		19.0 sq. ft
Boiler pressure		170 psi
Coupled wheels		5 ft 2 in.
Tender wheels		4 ft 2 in.
Tractive effort		21,875 lb.
Length over buffers		52 ft 6 in.†
Wheelbase	Locomotive	16 ft 3 in.
	Tender	13 ft 0 in.
	Total	38 ft 10 in.
Weight in working order	Locomotive	50 tons 10 cwt
	Tender	43 tons 2 cwt
	Total	93 tons 12 cwt
Max. axle loading		18 tons 0 cwt
Water capacity		3,500 gallons
Coal capacity		6 tons 10 cwt

* With superheater, † Engine and tender.

Between the years 1889 and 1893 Patrick Stirling introduced into service 25 0-4-4Ts with 5 ft 8 in. driving wheels, to supplement the motive power working the GNR London suburban services. Known as the 'Metropolitan' tanks the locomotives were classified 'G1' by both the GNR and LNER. After the introduction of Ivatt's 'C2' class, later LNER 'C12' class, 4-4-2Ts on the London suburban services a few 'G1s' were sent to country depots, and from 1907 when the Gresley 0-6-2Ts were appearing on the scene, the remainder of the 0-4-4Ts were transferred away. During this period and before the arrival of the 'C2s' at New England, a 'G1' tank engine was occasionally allocated to the Ramsey branch passenger duties. Even after World War I when the 'C2' class 4-4-2Ts enjoyed an almost near monopoly on the Huntingdonshire branch, 'G1' Nos. 939 and 943 made occasional forays to Ramsey. Their stay, however, was shortlived for both were withdrawn from traffic for scrapping in June 1924 before receiving their allocated LNER numbers 3939 and 3943 respectively. The leading dimensions of the 'G1' class 0-4-4Ts were:

Cylinders	2 inside	18 in. x 26 in.
Motion		Stephenson with slide valves
Boiler	Max. diameter	4 ft 6 in.
	Barrel length	10 ft 1 in.
	Firebox	5 ft 6 in.
Heating surface	Tubes 213 x 1¾ in.	1,016.0 sq. ft
	Firebox	103.0 sq. ft
	Total	1,119.0 sq. ft
Firebox grate area		16.25 sq. ft
Boiler pressure		160 psi
Coupled wheels		5 ft 8 in.
Trailing wheels		3 ft 2 in.
Tractive effort		16,846 lb.
Length over buffers		33 ft 5½ in.
Wheelbase		22 ft 6 in.
Weight in working order		53 tons 9 cwt
Max. axle loading		17 tons 10 cwt
Water capacity		1,000 gallons
Coal capacity		3 tons 0 cwt

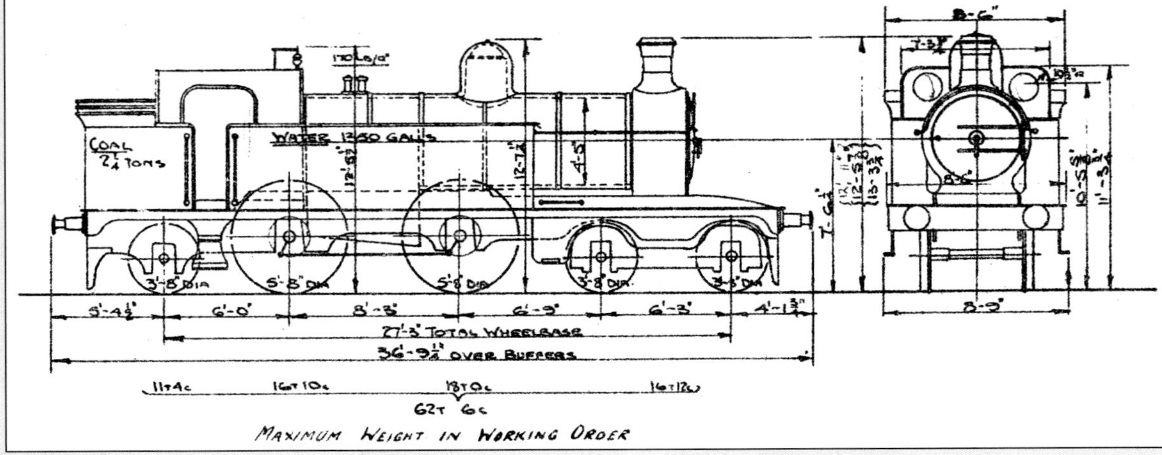

GNR class 'C2' 4-4-2Ts, later reclassified to 'C12' by the LNER were responsible for working the branch passenger services after their displacement from London suburban services until the full withdrawal of the passenger timetable in 1931.

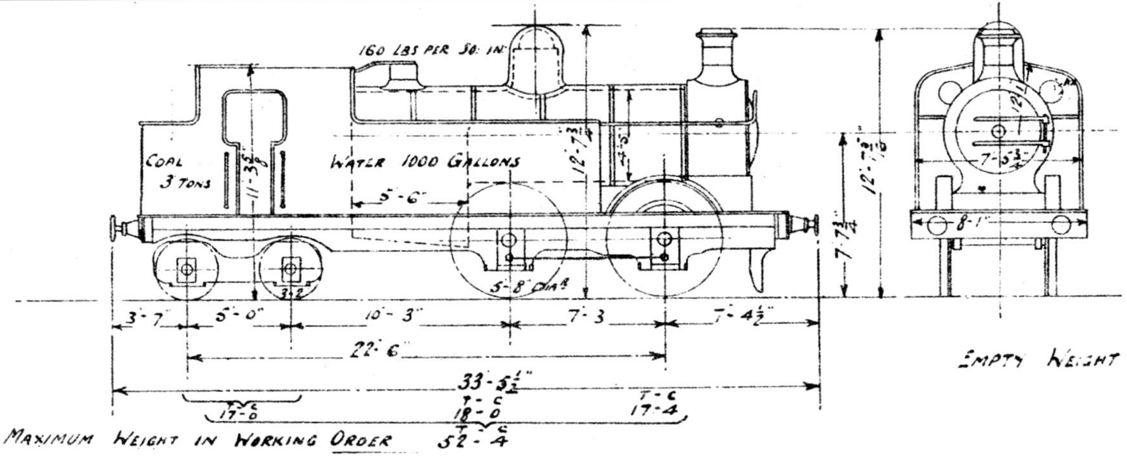

GNR class 'G1' 0-4-4Ts were occasionally used on the Ramsey branch passenger services in their final years.

LOCOMOTIVES AND ROLLING STOCK 145

GNR 'C2' class 4-4-2T No. 1508 taking water at Holme in the early 1920s. The suspension footbridge was provided in 1879 after the death of a schoolboy on the adjacent level crossing. It was removed in the late-1920s and only replaced in 1936. Note also the elderly four-wheel tender stabled in the siding alongside the Ramsey branch. *Author's Collection*

The only class of tank locomotive to work regularly on the Ramsey services was Ivatt's 'C2' class 4-4-2s, later classified 'C12 'by the LNER. Originally introduced between 1898 and 1907 for use in the West Riding of Yorkshire and on the London suburban services, they were later superseded by Gresley's 'N1' class and 'N2' class 0-6-2Ts. Soon after World War I, most of the London-based locomotives were sent to country districts for use on branch line services and the Nottingham district suburban trains. New England shed received an allocation of six engines, and of these two were employed as Peterborough station pilots at the north and south end of the station. A further two were outbased at Stamford for the Essendine and Wansford branches, whilst another worked the Ramsey branch. The Ramsey locomotive handled both passenger and freight traffic for over a decade, but ceased with the withdrawal of the full passenger service in February 1931. Locomotives known to have been associated with the Ramsey North branch included:

GNR No.	LNER 1924 No.	LNER 1946 No.	BR No.	Withdrawn
1502	4502	7360	67360	January 1955
1505	4505	7363	67363	November 1958
1507	4507	7365	67365	May 1958
1508	4508	7366	67366	April 1957
1516	4516	–	–	November 1937
1521	4521	7376	67376	May 1958
1522	4522	–	–	June 1937
1527	4527	7380	67380	May 1958

No. 67380 worked the Railway Enthusiasts' Club 'Charnwood Forester' Railtour train across the branch in April 1957.

The principal dimensions of the class were:

Cylinders	2 inside	18 in. x 26 in.
Motion		Stephenson with slide valves
Boiler	Max. diameter	4 ft 6 in.
	Barrel length	10 ft 1 in.
	Firebox	5 ft 6 in.
Heating surface	Tubes 213 x 1¾ in	1,016.0 sq. ft
	Firebox	103.0 sq. ft
	Total	1,119.0 sq. ft
Firebox grate area		16.25 sq. ft
Boiler pressure		170 psi
Leading wheels		3 ft 8 in.
Coupled wheels		5 ft 8 in.
Trailing wheels		3 ft 8 in.
Tractive effort		17,900 lb.
Length over buffers		36 ft 9¼ in.
Wheelbase		27 ft 3 in.
Weight in working order		62 tons 0 cwt
Max. axle loading		18 tons 0 cwt
Water capacity		1,350 gallons
Coal capacity		2 tons 5 cwt

Ironically the only ex-GER class of locomotive to work across the GER Ramsey branch was Worsdell's 'Y14' class 0-6-0s, later designated class 'J15' by the LNER. Introduced in 1883, such was the success of the class that building continued until 1913 when all except 19 of the class of 289 engines were built at Stratford, with the others built by Sharp, Stewart & Co. The ubiquitous class worked on all lines of the former GER on express freight, pick-up goods and for those fitted with the Westinghouse brake, branch and excursion passenger trains. After the Grouping seven were rendered surplus on their home territory and were allocated to New England shed in the mid-1930s for use on the Whittlesea yard brick trip workings. It was during this period that the Ramsey North line received a visit from the class on engineers' specials but, as afar as can be ascertained, they were never utilized on the freight diagram which continued to be entrusted to the ex-GNR 0-6-0s.

During the 1955 strike by members of the Associated Society of Locomotive Engineers and Firemen, New England shed was virtually at a standstill and no trains ran to Ramsey North for some days. The railway authorities were desperate to dispatch essential traffic from both Ramsey North and St Mary's and devised a plan to send the locomotive outbased at the former GER shed at Huntingdon East, and manned by a National Union of Railwaymen footplate crew, to clear the branch of traffic. It was on these occasions that the line again played host to a 'J15' class 0-6-0, but as the Huntingdon crew were unfamiliar with the branch, goods foreman Marshall of Ramsey acted as pilotman from the junction to the terminus and back. On one trip, having arrived and formed its train at Ramsey North, it was discovered that the engine was short of water. It was thought politically undesirable to return to Holme to top up the tender

LOCOMOTIVES AND ROLLING STOCK

tank, and so local emergency measures were taken with the tank being replenished from the station domestic water supply. Needless to say this laborious process took some time to accomplish and it was over two hours later before the class 'J15' locomotive set out for Holme, and ultimately Huntingdon, with her important train. The engine then returned light to Huntingdon East shed.

The leading dimensions of the 'J15' class were:

Cylinders	2 inside	17½ in. x 24 in.
Motion		Stephenson with slide valves
Boiler	Max. diameter	4 ft 4 in.
	Barrel length	10 ft 0 in.
	Firebox	6 ft 0 in.
Heating surface	Tubes 242 x 1⅝ in.	1,063.8 sq. ft
	Firebox	105.5 sq. ft
	Total	1,169.3 sq. ft
Firebox grate area		17.9 sq. ft
Boiler pressure		160 psi
Coupled wheels		4 ft 11 in.
Tractive effort		16,942 lb.
Length over buffers		47 ft 3 in.*
Wheelbase	Locomotive	16 ft 1 in
	Tender	12 ft 0 in
	Total	35 ft 2 in.
Weight in working order	Locomotive	37 tons 2 cwt
	Tender	30 tons 13 cwt
Max. axle loading		13 tons 10 cwt
Water capacity		2,640 gallons
Coal capacity		5 tons 0 cwt

* Engine and tender

Another class associated with the Ramsey North line during the final years of steam was the ex-London Midland 2-6-0s designed by H.G. Ivatt, son of the former GNR locomotive engineer. As early as 1950 New England had an allocation of 20 of the class but they usually worked on the former Midland & Great Northern section to South Lynn, Cromer and Yarmouth Beach until the line closed in 1959. The engines formerly allocated to Yarmouth Beach, Melton Constable and South Lynn sheds were then dispersed throughout the Eastern Region and New England received an additional allocation. The LM '4MTs' quickly displaced the 'J6' 0-6-0s on many duties, and from then on a member of the class worked the branch until the demise of steam traction. The additional power proved popular with local enginemen, as did the enclosed cab and tender cab, which gave crews greater protection from the elements during adverse weather on the exposed sections of the branch. Locomotives known to have worked on the line included Nos. 43067, 43081, 43082, 43084, 43088, 43089, 43094, 43146, 43151 and 43153. Despite their popularity with locomotive crews, the 'J6' 0-6-0s and LM '4MT' 2-6-0s were unpopular with the local permanent way inspector who considered they were too heavy for the two timber bridges on the branch. The principal dimensions of the LM '4MT' class were:

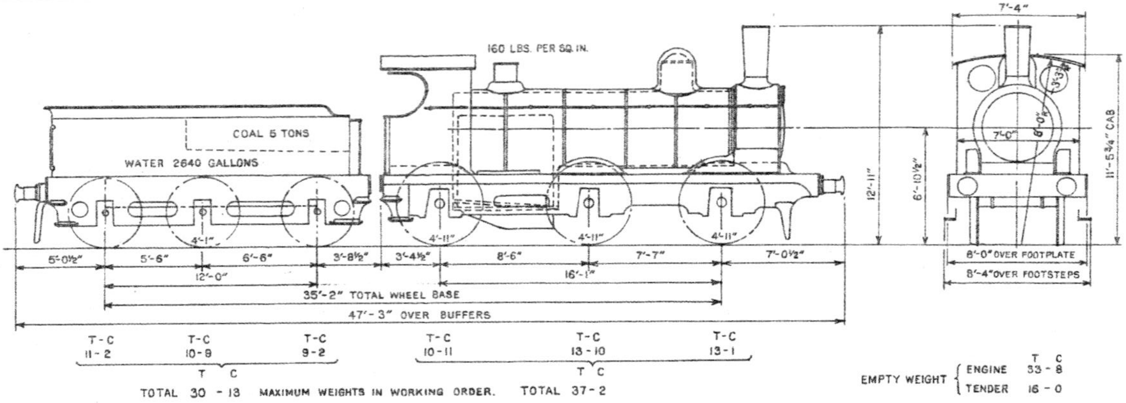

GER 'Y14' class 0-6-0, later reclassified 'J15' by the LNER, visited the Ramsey North branch in the 1930s and again during the ASLEF strike in 1955.

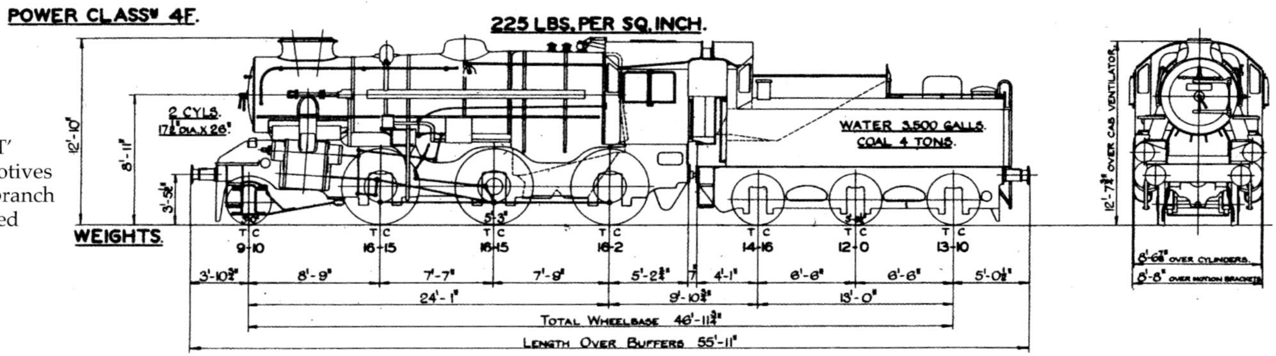

LM Ivatt '4MT' 2-6-0. Locomotives working the branch were only fitted with a single chimney.

Cylinders	2 inside	17½ in. x 26 in.
Motion		Walshaerts valve gear
Boiler	Max. diameter	5 ft 3 in.
	Barrel length	10 ft 10⅜ in.
	Firebox	7 ft 6 in.
Heating surface	Tubes 160 x 1⅝ in.	1,091.0 sq. ft
	Firebox	131.0 sq. ft
	Total	1,221.0 sq. ft
Superheater	24 x 5⅛ in.	231.0 sq. ft
Firebox grate area		23.0 sq. ft
Boiler pressure		225 psi
Leading wheels		3 ft 0 in.
Coupled wheels		5 ft 3 in.
Tractive effort		24,172 lb.
Length over buffers		55 ft 11 in.*
Wheelbase	Locomotive	24 ft 1 in.
	Tender	13 ft 0 in.
	Total	46 ft 11 ¾ in.
Weight in working order	Locomotive	59 tons 2 cwt
	Tender	40 tons 6 cwt
Max. axle loading		16 tons 15 cwt
Water capacity		3,500 gallons
Coal capacity		4 tons 0 cwt

* Engine and tender.

With the withdrawal of steam from the Eastern Region south of Peterborough in 1963, British Rail continued to classify the line RA4 and found the class '3/1', later class '08', 0-6-0 diesel shunting locomotives were adequate motive power for the Ramsey branch goods and they remained in charge until the withdrawal of freight facilities and closure of the line on and from 2nd July, 1973. All of the class '08' diesel-electric shunting locomotives numbered in the series D3445 to D3453, D3485 to D3489 and D3629 and D3630 allocated to New England, which worked the Ramsey North branch, were fitted with GEC equipment and had Blackstone engines. However in March 1972 No. D3527 with English Electric equipment was noted on the branch service. Other locomotives known to have ventured to Ramsey North included D4076 and D4077 and D4083 and D4084. The principal dimensions of the Blackstone 350 hp locomotives were:

Type		0-6-0
Weight in working order		48 tons
Tractive effort		35,000 lb.
Wheelbase		11 ft 6 in.
Wheel diameter		4 ft 6 in.
Width overall		8 ft 10 in.
Length overall		29 ft 3 in.
Height overall		12 ft 8⅝ in.
Min. curve negotiable		3 chains
Max. speed		20 mph
Fuel tanks	Main	585 gallons
	Service	83 gallons
	Total	668 gallons

THE RAMSEY NORTH BRANCH

LM class '4MT' 2-6-0 No. 43067 standing at Ramsey North with the branch freight train formed of just a goods brake van in the latter days of steam working. *The late Dr Ian C. Allen*

BR 350 hp diesel-electric shunting locomotive, later class '08'.

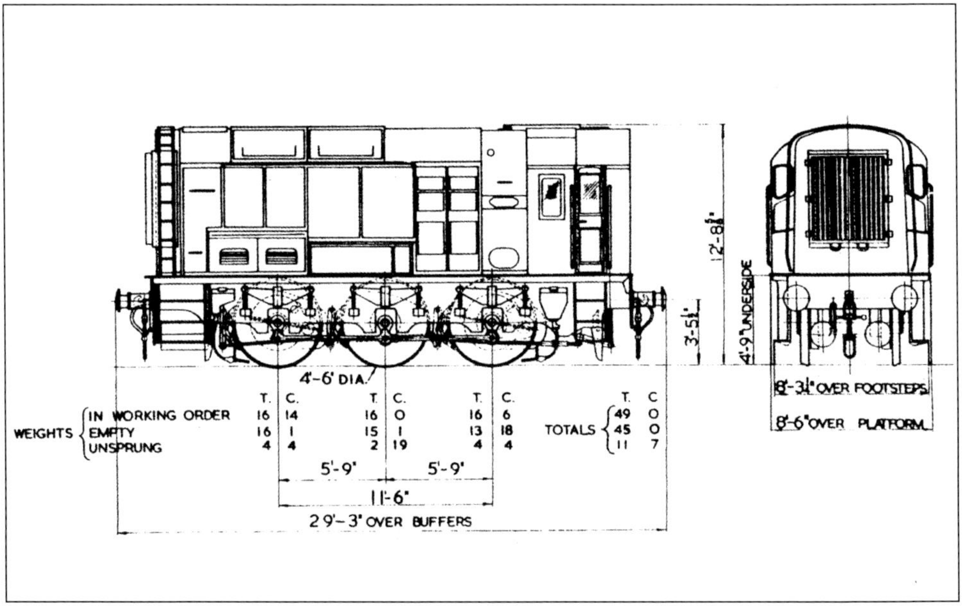

LOCOMOTIVES AND ROLLING STOCK

Lubricating oil sump		45 gallons
Cooling water	Radiator	90 gallons
	Engine	50 gallons
	Total	140 gallons
Brakes		Compressed air and hand brake on the locomotive. Vacuum brake equipment provided to work fitted stock and give proportional air braking on the locomotive when so working.
Engine		6 cylinder diesel Blackstone ER 6T type 350 hp at 750 rpm.
Main generator		DC self-ventilated GEC WT 821 type. 1 hr rating 228 kw, 600 amps, 380 volts at 750 rpm.
Auxiliary generator		DC self ventilated GEC type WT 700. 1 hr rating. 9.2 kw, 92 volts, 100 amps 2,520 rpm.
Traction motors		DC force-ventilated 4 pole GEC type WT 360, 1 hr rating 2 x 152hp, 360 volts, 370 amps at 800 rpm.
Air compressor		Westinghouse DH 16, motor driven.
Exhausters		Westinghouse 3V72, motor driven.

Facilities and staff

Motive power for the branch was supplied from Peterborough Station engine shed but about 1870 New England shed took over the responsibility for providing locomotives. Initially coded 35A by British Railways, the depot later became 34E. When New England closed on and from 30th September, 1969 the class '08' 0-6-0 diesel-electric shunting locomotives working the branch services were provided by the former GER March shed coded 31B. The locomotive out-based at Ramsey to work the branch was stabled in a small corrugated iron, timber-framed engine shed, measuring 45 ft by 15 ft and large enough to accommodate a small tender locomotive. The building was located on the up or south side of the main single line, west of the station and adjacent to the water tank. Water for the locomotive was obtained from a crane fed from this storage tank raised above the pump house, the latter also serving as a store. From the mid-1880s the working of the branch came under review and to achieve economies, the GNR authorities decided in 1885 to withdraw the out-based branch locomotive and work the line on an out and back basis from New England shed. The condition of the engine shed subsequently deteriorated and in 1896 the GNR wrote to the GER asking for a decision as to the future use of the building. The GER General Manager reported the matter to the Traffic Committee on 10th February, 1897, and after due discussion they authorized the demolition of the structure, leaving the tank house as sole evidence of the existence of the shed.

It was customary to top up the tanks of the locomotives at both Ramsey and Holme but after rationalization the water supply, except for domestic purposes, was withdrawn from Ramsey and all water was taken at Holme. A small coke, later coal-stage was originally provided at Ramsey as the branch locomotive was coaled during the day, as and when required, by the fireman. A cleaner was provided on nights to clean the depot and the locomotive, as well as offloading coal from wagons to the coal stage. At Holme the water column was located 28 yards south of the up side platform and served the up main line and the Ramsey branch. It was fed from a 15,000 gallons tank supplied from the local Drain. Consequently the water was very bad but was softened at a high cost by adding chemicals. The estimated consumption of water in 1901 was three million gallons, which by 1906 had risen to 4.5 million gallons. At both Holme and Ramsey the station domestic coal supply was used to heat the locomotive water column feed pipes to prevent freezing up during winter weather.

```
NEW   ENGLAND   DEPOT—Continued.
                                              WEEK DAYS.
                    No. 15.
     arr.                       dep.
     a.m.                       a.m.
    ............On Duty............  3 55
    ............Loco ................  4 40 L
    ............Peterboro'..........  4 55 (106) G
  5 18 ..... Holme ................  5 58 G
  6 30 ..... Ramsey North......  6 55
  7 14 ..... Holme ................  7 43
  7 59 ..... Ramsey North......  9  5
  9 19 ..... Holme.................  9 30
  9 43 ..... Ramsey North......  9 48
 10  2 ..... Holme ................ 10 15
                                  p.m.
 10 33 ..... Ramsey North...... 12 45
  p.m.
 12 59 ..... Holme.................  1 35
  1 54 ..... Ramsey North......  2 15 G S O
  2 57 ..... Holme.................  3  5 L S O
  3 18 ..... Ramsey North......  3 40
  3 54 ..... Holme.................  4 15
  4 29 ..... Ramsey North......  4 45 G S X
  5 27 ..... Holme.................  5 37 L S X
  5 50 ..... Ramsey North......  6 15
  6 29 ..... Holme.................  6 38
  6 52 ..... Ramsey North......  7 15
  7 29 ..... Holme.................  8  5
  8 19 ..... Ramsey North......  8 30 L
  8 45 ..... Holme.................  9  5 L
  9 25 ..... New England ......     L
    ............Loco ................
      Men change 10.2 a.m. and 3.58 p.m.
```

New England depot locomotive and enginemen's workings
on the Ramsey North branch, July 1925.
(*Key:* G – Goods; L – Light locomotive)

When the line first opened to traffic only one set of footplate staff was employed at Ramsey to cover the full hours of working, as well as preparation and disposal of the locomotive. Later a cleaner was also employed. With the rearranged working and the withdrawal of the out-based locomotive from the branch, all men were transferred to New England shed. Once the Peterborough depot operated the branch, two sets of men were required, but later three sets were provided to cover the branch services, the first set signing on at 3.55 am before being relieved at Holme at 10.02 am by the second set who had travelled out passenger from Peterborough. The third set relieved the second set at Holme at 3.58 pm and worked the remainder of the trains on the branch before returning with the light engine from Holme to Peterborough at 9.05 and arriving at New England shed at 9.25 pm.

The locomotive foreman or driver-in-charge during the early years received half a day's pay for administrative duties, which included the submission of duty tickets and coal and oil returns to the District Office at New England shed. Men out-based at Ramsey only worked the branch services and occasional light engine movements to and from Peterborough. In later years only New England men 'signed the road' between Holme and Ramsey North.

The locomotive outbased to work the Ramsey branch rarely travelled beyond Holme unless sent for changeover with a relieving engine, which usually arrived with the first down goods train from Peterborough on a Monday morning. The new locomotive then worked the branch for up to two weeks before changeover. After the closure of Ramsey shed, the locomotive allocated to work the Ramsey line hauled the early morning up freight from New England, Spital or Yaxley yard to Holme before working the 5.58 am goods train thence to Ramsey. It was then responsible for all the branch workings before returning to New England shed in the evening. Depending on the amount of freight to be moved from the branch stations, in the evening the locomotive either worked a goods train to New England or returned light engine to shed. With the absence of engine turntables at both Holme and Ramsey, the branch locomotive usually worked boiler first to Holme and tender or bunker first to Ramsey.

The engine headcode carried by the branch trains in GNR days was one lamp on the chimney and one lamp over the right-hand buffer for ordinary passenger, excursion and empty coaching stock trains. Goods, mineral and ballast trains carried one headlamp over the left-hand buffer, whilst light engines carried a lamp over the right-hand buffer. The headlamps carried were illuminated during the hours of darkness or during fog or falling snow. It is interesting to note that most GNR engines actually carried the lamp bracket on the front of the chimney and not at the top of the smokebox. As boilers increased in diameter and the chimney became shorter, the lamp bracket was resited to the conventional position on top of the smokebox. After Grouping the LNER adopted the standard stopping passenger train code of one white light under the chimney, whilst freight trains carried the appropriate class headcode.

The following whistle codes were applicable to the branch when under GNR control:

1877 Holme	Main line	1 whistle
	Through road	1 long whistle
	Outlet from coal drop siding	1 long whistle
	Down siding to down main line	2 whistles
	Outlet from down shunting siding	3 whistles
	Outlet from up shunting siding	3 whistles
1912 Holme	Branch line to branch siding	1 long 1 short
	Branch siding to wharf	1 long 2 short
	Branch line and No. 1 up siding	2
	Up main and wharf siding	2 short
	Up main and No. 1 up siding	2 crows
	Up main and No. 2 down siding	1 long 3 short
	Up main and No. 3 down siding	1 long 2 short
	Down main and No. 1 up siding	1 long
	Down main and No. 1 down siding north	1 long 1 crow
	Down main and No. 2 down siding north	2 long 2 crows
	Down main and No. 1 down siding south	1 crow
	Down main and No. 2 down siding south	1 crow 1
	Down main and No. 3 down siding south	1 crow 2
	Down sidings 2 & 3 to down main south	1 crow 1
	Up siding 2 to up siding 1	3
Ramsey Ground Frame	Run round road and single line	4
Ramsey	Running line and run round loop	2
	Running line and down sidings	3
	Running line and up sidings	5
Whittlesea Mere Level Crossing	on approach	1 long

From 1927 the LNER rationalized the number of whistle codes used on the branch, and these remained in use until closure:

Holme Station Box	Branch and branch siding	1 long 1 short
	Branch and No. 1 up siding	2 short
Ramsey North (ground frame)	Run round road and single line	4 short
Ramsey North Station Box	Main line and run round road	2 short
	Main line and down sidings	3 short
	Main line and up sidings	5 short

In the event of a mishap or derailment, Peterborough North and later New England breakdown vans or crane were employed to clear the line. A 35 ton capacity steam crane, No. 343A built by Cravens, was used in the early years of the 20th century, but latterly the 45 tons capacity steam crane No. 941599, later renumbered by British Railways to 110 and then 330110, was utilized. It was built by Cowans, Sheldon & Co. Ltd of Carlisle (Works No. 6871) and delivered in 1940. The timber wagon re-railing ramps available at Ramsey and Holme were not on any account to be used for rerailing locomotives, although unofficially on the rare occasions when engines became derailed by one pair of wheels, these ramps were found most useful. Ropes were also used to assist in the rerailing of engines.

It was often the practice for footplate crews to leave the branch locomotive by the buffer stops on granary or outside road at Ramsey North after completion of shunting duties, in order to gain speedy access to the neighbouring Railway Hotel for liquid refreshment. The siding was, however, situated on a slightly rising gradient towards the buffer stops, a generally unknown factor which became all too evident to at least two footplate crews. On the first occasion in the early 1930s, the driver and fireman of a 'J4' class 0-6-0 duly stabled their steed in the siding and made off for the hostelry. Unfortunately in his haste to get a pint and thinking the locomotive was on a level gradient, the fireman omitted to screw the handbrake down on the tender. The driver equally intent on quenching his thirst left the reversing lever one notch in forward gear instead of mid-gear. The combination of errors and a blowing regulator gland resulted in an incident that would have credited a Will Hay comedy. With the intrepid crew comfortably propping up the bar, the 'J4' 0-6-0 started to move off very slowly of its own accord. The advantage of the few yards of slightly falling gradient, gave the locomotive the impetus to make off for Holme. Luckily the porter-signalman on duty was locking the signal box door as the locomotive wheezed past. Perturbed to find the engine moving when all signals were at danger, he was dumbfounded to find the cab devoid of driver and fireman. Realizing where the footplatemen were ensconced, he raced for the hotel to sound the alarm, as the engine hissing gently trundled at walking pace towards St Mary's. In the hostelry pints and pies were quickly discarded, as the crew made off across the goods yard after their wayward steed. The fireman being the more agile of the pair, grabbed the porter's bicycle that happened to be handy and pedalled furiously along the cess beside the permanent way, sometimes falling or running over awkward sections of ballast. Half a mile beyond the outlet points the locomotive was brought to a stand. The cycle was loaded aboard and the fireman reversed the runaway back towards the station, picking up the exhausted driver on the way. Both men were much the wiser as to their future actions when stabling locomotives at Ramsey, and were indeed out of pocket over the escapade. Although the authorities were unaware of the event, the porter who owned the bicycle was all too aware that both wheels of his machine were buckled as it was pedalled along the trackside. Driver and fireman graciously paid for the replacement wheels.

Ironically many years later, whilst a class '08' diesel-electric shunting locomotive was stabled in the same siding for the same reason, the footplate crew failed to screw down the handbrake. Although not under power the falling gradient from the buffer stops gave the locomotive the impetus to move of its own accord. Once again the bliss of a pie and a pint was rudely shattered, but fortunately the wayward diesel had only just passed the outlet points to the station when the driver managed to scramble aboard and apply the brake. Thereafter immaculate shunting was performed at Ramsey for the rest of the day.

Over the years the various level crossing gates on the branch succumbed to the advances of moving trains on a number of occasions, leaving the civil engineering staff to repair or replace the splintered woodwork. The demolition process usually resulted as a direct result of the negligence by crossing keepers who failed to open their gates in time for the approaching train or drivers who failed to take note of the gate distant signals or more importantly the position of the gates in relation to the railway. Probably the most amusing incident involved the demolition of Whittlesea Mere level crossing twice in one week, through the negligence of the same driver. The second occasion was all the more galling as the civil engineering staff had only just finished replacing the gates after the first incident. When questioned at the subsequent inquiry over his action and lack of attention to the road ahead, the driver explained to the motive power inspector that at the time he was rolling a cigarette, but when trying to light it had dropped his box of matches. He was in the process of picking up the loose matches from the footplate when the sound of splintering woodwork was heard from the front of the locomotive. On being cross-examined as to his actions, the fireman explained he was also preoccupied picking up the matches and both he and the driver had forgotten the close proximity of the gates, which they expected to be open for the train. The inspector was not the least impressed by the story, but as the crossing was in the charge of a relief gatekeeper on both occasions when the collisions occurred, he suspended both the driver and the fireman from duty for two days without pay.

Coaching stock

Unlike locomotives, the GNR placed no weight or loading gauge restrictions on the rolling stock used on the Ramsey line and conventional branch line rolling stock was utilized. Early in GNR days the number of coaches permitted on passenger trains on the Holme to Ramsey line was not to exceed 20 vehicles. When the trains were composed of not more than six coaches, one brake van was included in the formation. For six coaches and not exceeding 12, two brake vans were required whilst a train formed of over 12 vehicles required three brake vans. By the turn of the century the maximum number of vehicles on the branch passenger train was restricted to 20 coaches with empty coaching stock permitted to increase to 25, although it is doubtful if trains of these lengths were ever conveyed.

By 1905 the following number of vehicles and brake power had to be provided on passenger services:

Max. number of vehicles	Passenger train	20
Max. number of vehicles	Empty coaches	25

When continuous brake from locomotive not available:

Not exceeding 9 vehicles	1 brake van
Not exceeding 18 vehicles	2 brake vans
Not exceeding 20 vehicles	3 brake vans

When continuous brake worked from the locomotive was available on at least half the train:

Not exceeding 10 vehicles   1 brake van
Not exceeding 20 vehicles   3 brake vans

Fully fitted train with continuous brake worked from the locomotive:

Not exceeding 12 vehicles   1 brake van
Not exceeding 20 vehicles   2 brake vans

The following additional tail loads were authorized on the Holme to Ramsey line:

Four carriages conveying passengers and not more than four horseboxes, carriage trucks or vans fitted with automatic brakes complete may, if necessary, be attached behind the rear brake van or carriage brake of passenger trains. Not exceeding one vehicle fitted with automatic brake pipes only may be attached behind the rear brake van or carriage brake of passenger trains.

From the opening of the line until the 1890s the coaching stock was exclusively four-wheel and provided with oil lighting. Trains were initially formed of up to 12 vehicles but this was soon reduced to eight vehicles formed of first class, composite first/second, full third class and brake/thirds. Full brakes were also provided in the absence of brake thirds. When the expected traffic failed to materialize the normal formation was reduced to three or four coaches except on market days.

As more modern vehicles became available in the King's Cross suburban area some of the displaced stock was sent to country areas. These included six-wheel vehicles, which after arriving in the Peterborough district soon infiltrated to the Ramsey branch. Trains were then formed of three or four vehicles including composite to diagram 154 or 156, two thirds to diagram 245 and brake third to diagram 281 or full brake to diagram 301 or 303. The leading dimensions of the these vehicles were

Diagram	*154*	*156*	*245*	*281*
	Six wheel	Six wheel	Six wheel	Six wheel
	Composite	Composite	Third	Brake third
Length over buffers	38 ft 6 in.	36 ft 3 in.	35 ft 6 in.	38 ft 3 in.
Length over body	35 ft 0½ in.	32 ft 10½ in.	32 ft 1½ in.	34 ft 10½ in.
Max. width	8 ft 8¼ in.	8 ft 8¼ in.	8 ft 8¼ in.	8 ft 10¾ in.*
Max. height	12 ft $4^1/_8$ in.	12 ft $4^1/_8$ in.	12 ft $4^1/_8$ in.	12 ft $4^1/_8$ in.
Wheelbase	24 ft 3 in.	23 ft 2 in.	22 ft 5 in.	24 ft 6 in.
Seating, first class	12	12	–	–
Seating, third class	30	30†	50	40
Weight	15 t. 6 cwt	14 t. 16 cwt	13 t. 16 cwt	14 t. 11 cwt

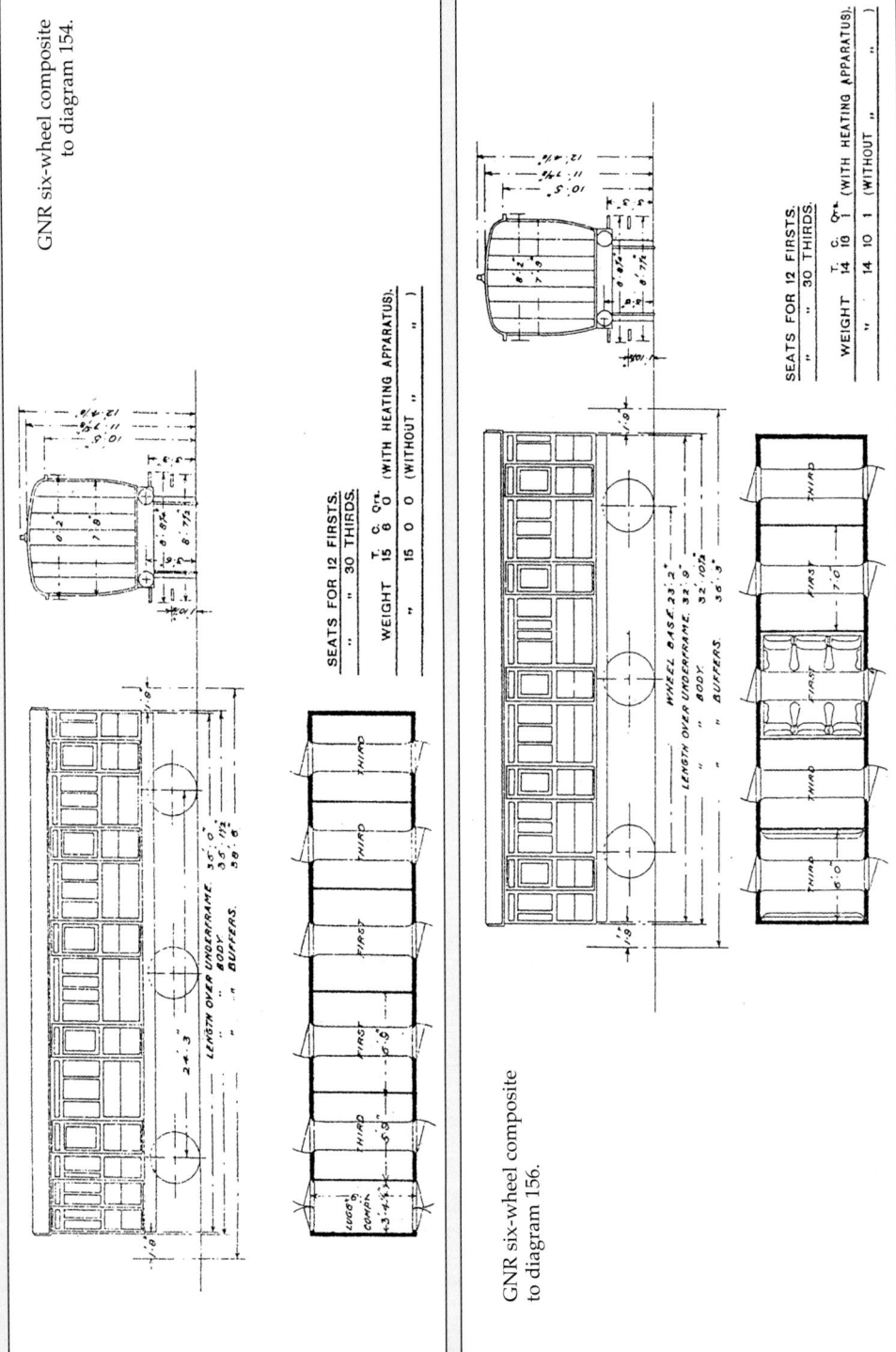

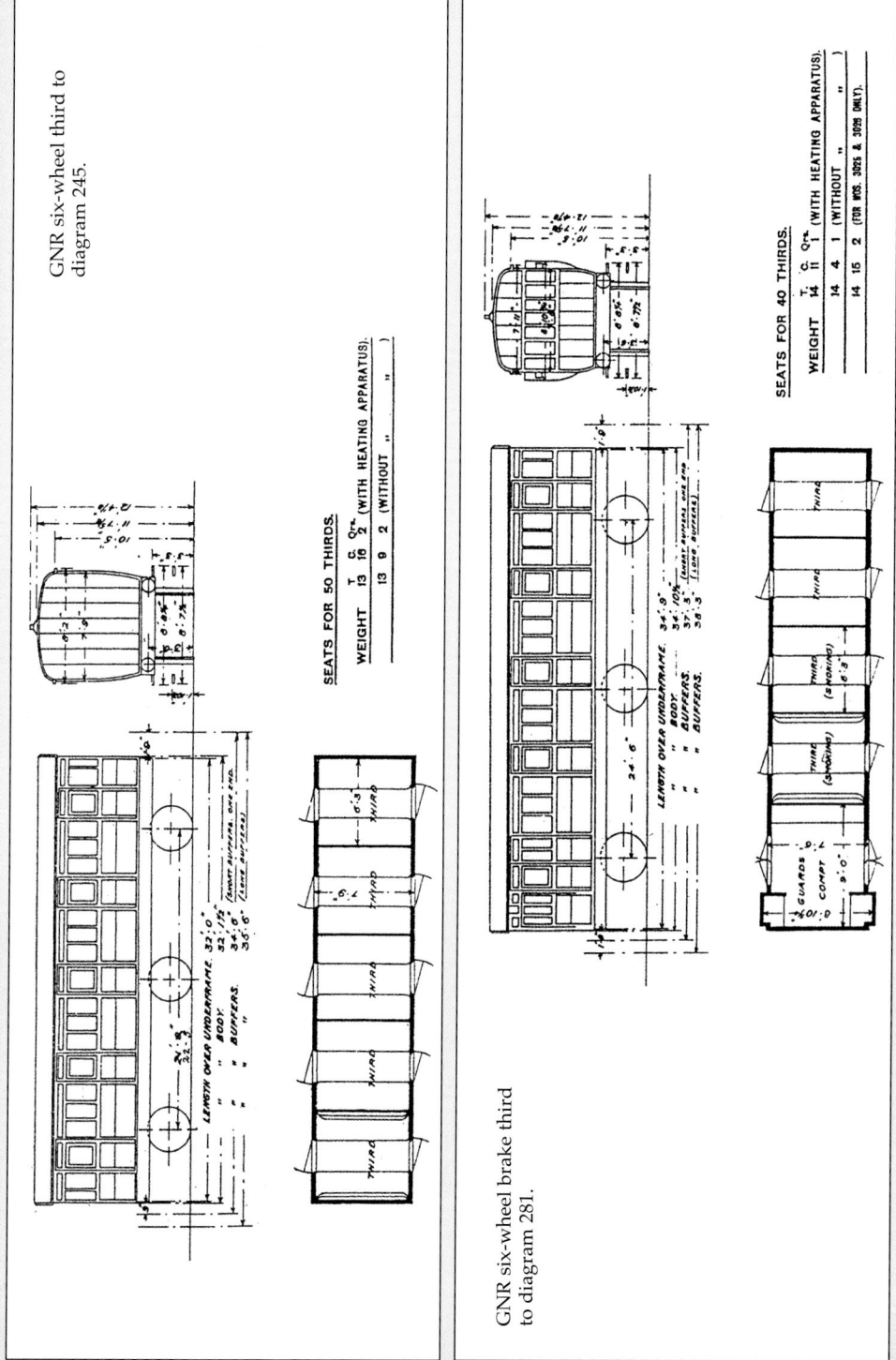

GNR six-wheel third to diagram 245.

GNR six-wheel brake third to diagram 281.

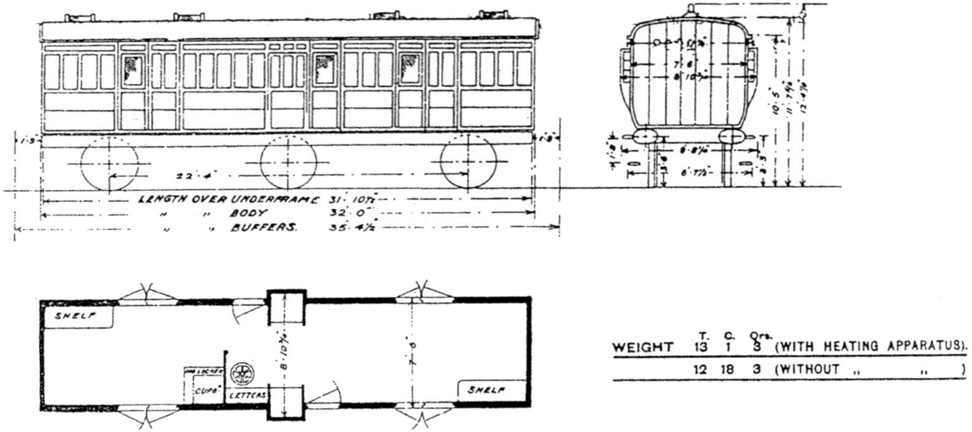

GNR six-wheel luggage brake van to diagram 301.

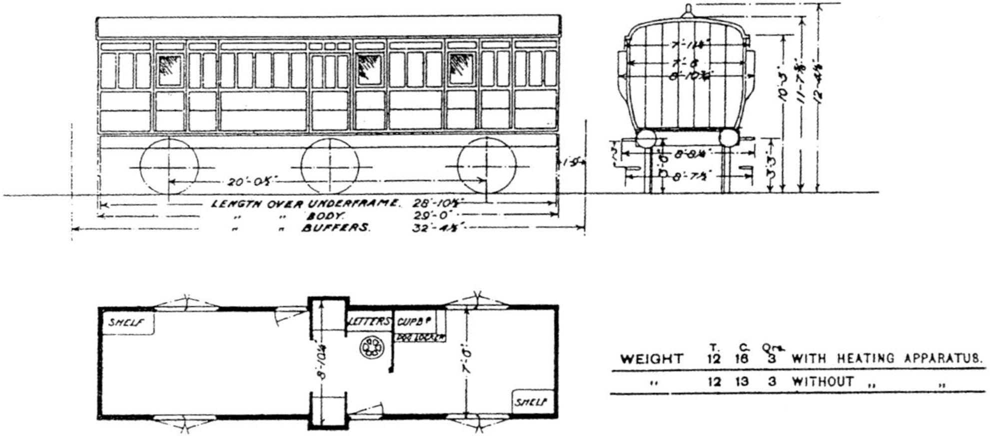

WEIGHT	T.	C.	Qrs.	
	13	1	3	(WITH HEATING APPARATUS.)
"	12	18	3	(WITHOUT " ")

GNR six-wheel luggage brake van to diagram 303.

WEIGHT	T.	C.	Qrs.	
	12	16	3	WITH HEATING APPARATUS.
"	12	13	3	WITHOUT " "

	Diagram	301	303
		Six wheel	Six wheel
		Full brake	Full brake
	Length over buffers	35 ft 4½ in.	32 ft 4½ in.
	Length over body	32 ft 0 in.	29 ft 0 in.
	Max. width	8 ft 10¾ in.*	8 ft 10¾ in.*
	Max. height	12 ft 4¹/₈ in.	12 ft 4¹/₈ in.
	Wheelbase	22 ft 4 in.	20 ft 0½ in.
	Weight	12 t. 19 cwt	12 t. 14 cwt

* Over guard's ducket. † Also second class.

Vans were often added to the passenger trains for conveyance of milk or perishable traffic.

During the latter part of the first decade of the 20th century many of the former six-wheel stock were converted to two and three articulated sets, which gave the vehicles a further lease of life. As well as working on the main line many were allocated for use on GNR branch lines. During this period the Holme to Ramsey branch was allocated a set and it is known that diagram 197 and 198 twin sets worked the services augmented by a diagram 218K triple articulated set. Other diagram vehicles might also have been utilized but the leading dimensions of the diagram 197, 198 and 218K are given below.

Diagram 197 Howlden Twin Articulated Set

	Brake/third		Composite
Length over buffers		75 ft 1¾ in.*	
Length over body	34 ft 10½ in.		35 ft 9¼ in.
Max. width	8 ft 8¼ in.		8 ft 8¼ in.
Max. width over duckets	8 ft 10¾ in.		–
Max. height	12 ft 4¹/₈ in.		12 ft 4¹/₈ in.
Wheelbase		64 ft 7¾ in.*	
Seating 1st class	–		10
3rd class	40		20
Weight in working order		31 tons 0 cwt*	

Diagram 198 Howlden Twin Articulated Set

	Brake/third		Composite
Length over buffers		73 ft 3½ in.*	
Length over body	34 ft 10½ in.		33 ft 11½ in.
Max. width	8 ft 7½ in.		8 ft 7½ in.
Max. width over duckets	8 ft 10¾ in.		–
Max. height	12 ft 4¹/₈ in.		12 ft 4¹/₈ in.
Wheelbase		62 ft 10¼ in.*	
Seating 1st class	–		12
3rd class	40		30
Weight in working order		31 tons 4 cwt 3 qtrs*	

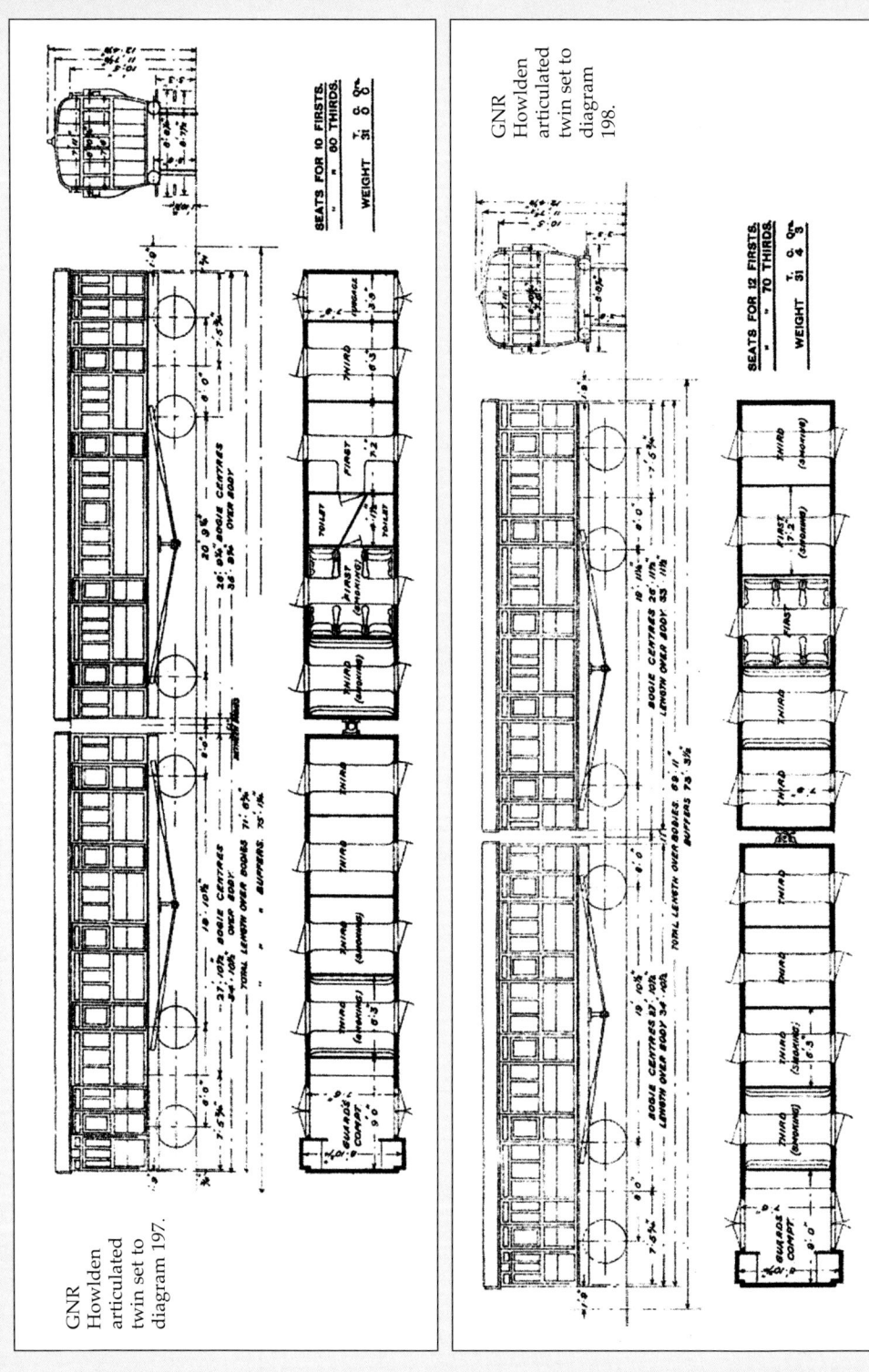

Diagram 218K Howlden Triple Articulated Set

	Brake/third	Third	Composite
Length over buffers		107 ft 10 in.*	
Length over body	34 ft 10 in.	32 ft 1½ in.	35 ft 3½ in.
Max. width	8 ft 8¼ in.	8 ft 8¼ in.	8 ft 8¼ in.
Max. width over duckets	8 ft 10¾ in.	–	–
Max. height	12 ft 4^{1}/$_8$ in.	12 ft 4^{1}/$_8$ in.	12 ft 4^{1}/$_8$ in.
Wheelbase		97 ft 9¾ in.*	
Seating 1st class	–	–	10
3rd class	30	50	20
Weight in working order		43 tons 14 cwt 2 qtrs*	

* Total length of set.

From 1931 the branch train was still formed of a three-set but later in the 1930s some two-coach articulated bogie suburban stock was introduced. In World War II a twin-set diagram 44 was allocated to work the passenger services. This set was formed of a non-corridor, non-lavatory brake composite, with two first and three third class compartments, coded BC and a non-corridor, non-lavatory full third, with four compartments, coded T weighing 32 tons and with accommodation for 10 first class and 60 third class passengers. The same set was used in 1946 and nearly until the withdrawal of passenger services. An additional non-corridor, non-lavatory brake third coded BT, was held as spare vehicle at Holme for working mixed trains. When passenger services were withdrawn in 1947 bogie coaches were in use. Trains were formed of Gresley bogie composite and bogie brake third but often, older vehicles of GER or GNR vintage were substituted. In World War II when movement of air force personnel was of priority, the train was often strengthened and was formed of four or five bogie coaches of various vintages.

When not in use and especially at night the branch coaching stock was usually stabled in the middle road at Ramsey. For a short period in the 1920s when the last up passenger working ran from Ramsey to St Mary's only, the vehicles were stabled in St Mary's goods yard to be picked up by the first down goods train the following day, or if a Saturday on the following Monday. At various times when the coaching stock was stabled at Holme, the vehicles were stored in the curved siding on the up side of the main single branch line. It is interesting to note that until 1947 the tail lamp used exclusively on the branch passenger train was of GNR design, painted red and overstamped 'LNER Ramsey North'.

Wagons

A short description of rolling stock provided by the GNR for the conveyance of goods traffic is appropriate. The details are not exhaustive but give guidance to the general use vehicles used on the branch trains. The wagons used by the GNR in the early years were wooden open vehicles with side doors and fitted with dumb buffers. Where grain, straw or merchandise was susceptible to wet weather, a tarpaulin sheet was utilized to protect the contents of the wagon. The brake van at the tail of the train would have been a 10 ton vehicle. From the

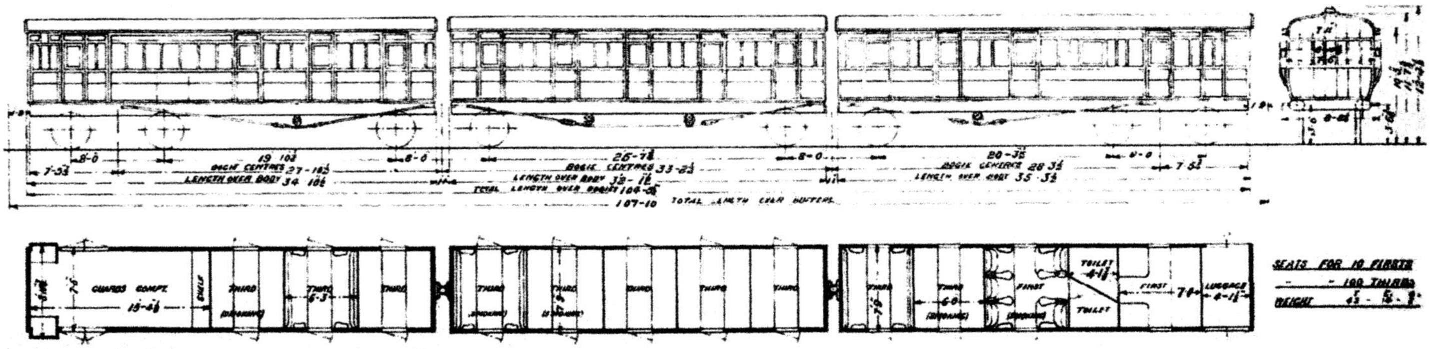

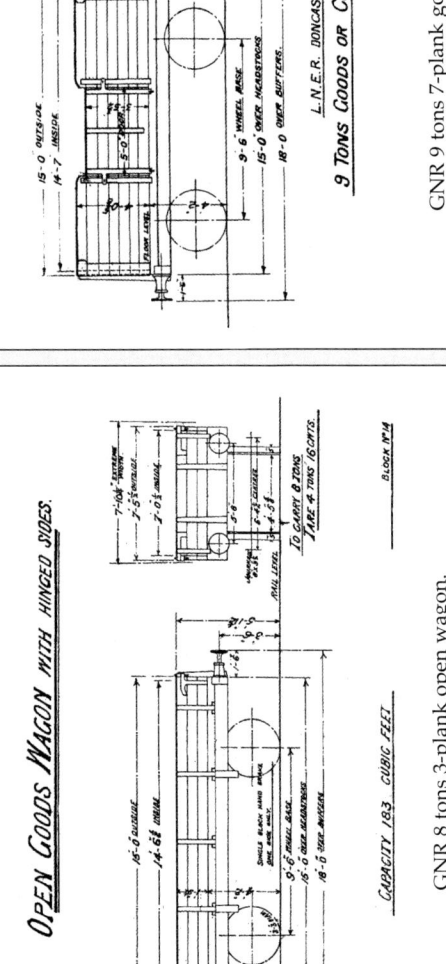

GNR Howlden articulated triplet set to diagram 218K.

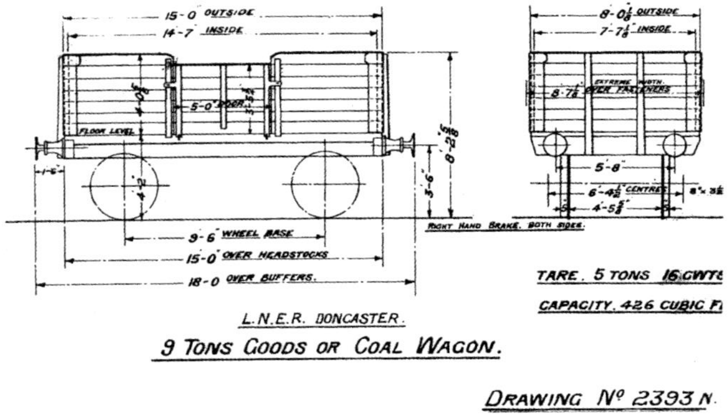

GNR 9 tons 7-plank goods or coal open wagon.

GNR 8 tons 3-plank open wagon.

1870s the GNR utilized 9 ton capacity four-plank bodied-open wagons with wooden frames, with 9 ft 6 in. wheelbase and length over buffers of 17 ft 8 in. for the conveyance of general merchandise and minerals. These wagons were gradually superseded by 9 ton four-plank opens measuring 15 ft 0 in. over headstocks and a wheelbase of 9 ft 6 inches, 9 ton capacity six-plank with 9 ft 6 in. wheelbase, also measuring 15 ft 0 in. over headstocks and 10 ton capacity opens with 10 ft 0 in. wheelbase and measuring 19 ft 0 in. over headstocks. Two varieties of lowside machine wagons would also have been used, both of 9 tons capacity, one with 9 ft 6 in. wheelbase and 18 ft 0 in. over headstocks and the other 10 ft 0 in. wheelbase and 19 ft 0 in. over headstocks. For fruit and perishable traffic, 6 ton and 8 ton capacity ventilated vans were provided, measuring 16 ft 0 in. over buffers. Later 8 tons capacity covered goods vans with 9 ft 10 in. wheelbase and overall height of 11 ft 4 in. were also utilized. Another variation of the same dimensions had a 10 ft 0 in. wheelbase. Cattle traffic to and from Ramsey would have brought several types of cattle wagons to the branch, including 6 tons capacity measuring 18 ft 2½ in. and later 6 tons capacity measuring 19 ft 0 in. Both were equipped with through vacuum pipes, screw couplings and oil-lubricated axleboxes. At the tail of the train could be found a four-wheel goods brake van of 10 ton, 13 ton, 15 ton or 20 ton capacity, all measuring 21 ft 6 in. over buffers, 18 ft 6 in. over headstocks and having 10 ft 0 in. wheelbase. In addition many wagons owned by other companies were used to deliver and collect agricultural and maltings traffic, whilst coal and coke supplies usually came in private owner coal wagons. These fell into two categories, those belonging to the collieries, and merchants and factors wagons, including those of the Peterborough Co-operative Society Ltd and Coote & Warren, which were loaded at the collieries. The GNR also operated a fleet of coal wagons and the following would have been seen on the Ramsey branch: 9 tons and 10 tons capacity, seven-plank, 9 tons capacity five-plank and six-plank, 10 tons five-plank and 15 tons capacity eight-plank, all with 9 ft 6 in. wheelbase and measuring 15 ft 0 in. over headstocks. Locomotive coal was delivered in 10 ton seven-plank open wagons, 15 tons capacity eight-plank, 12 tons capacity seven-plank and 20 tons capacity nine-plank opens, the latter with 12 ft 0 in. wheelbase and 21 ft 6 in. body length.

After Grouping the GNR wagons, continued in use but gradually LNER standard design wagons made an appearance. The most numerous were probably the 12 ton, five-plank opens with 8 ft 0 in. wheelbase to code 2, and 12 ton, six-plank opens with 10 ft 0 in. wheelbase to code 91 built after 1932. Later types included a 13 ton seven-plank open wagon to code 162 measuring 16 ft 6

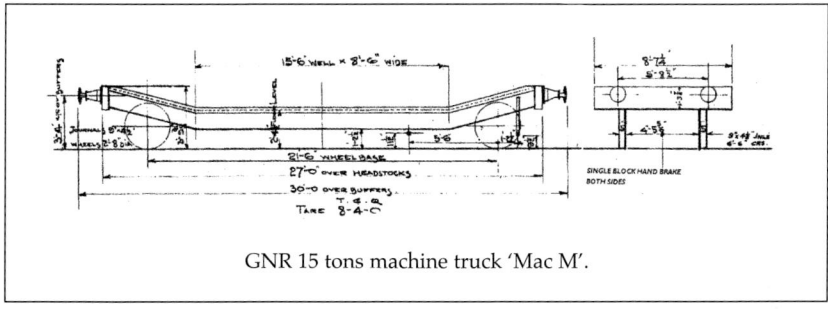

GNR 15 tons machine truck 'Mac M'.

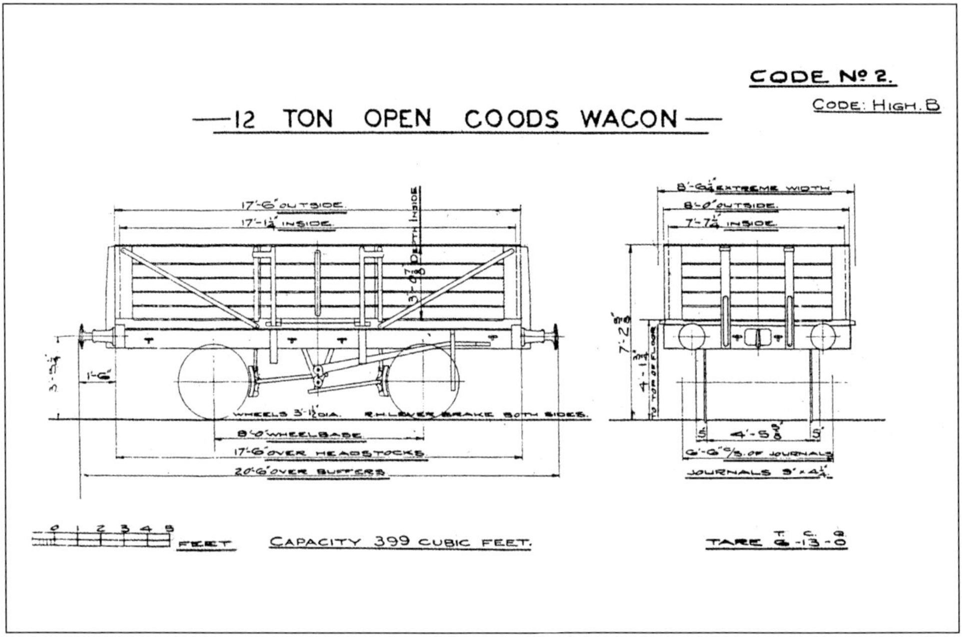

LNER 12 tons open goods wagon to code 2.

LNER 13 tons open goods wagon to code 162.

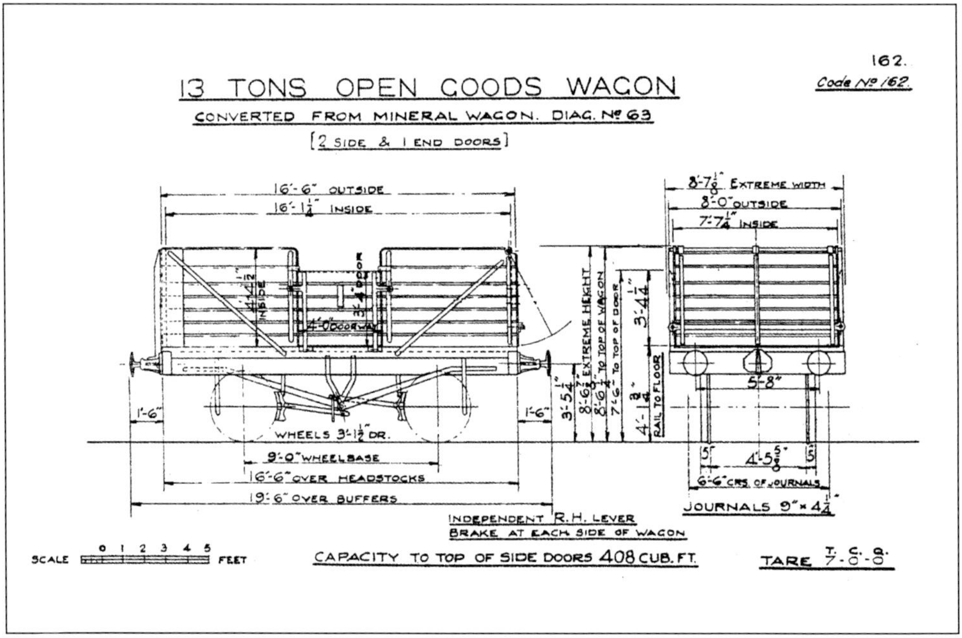

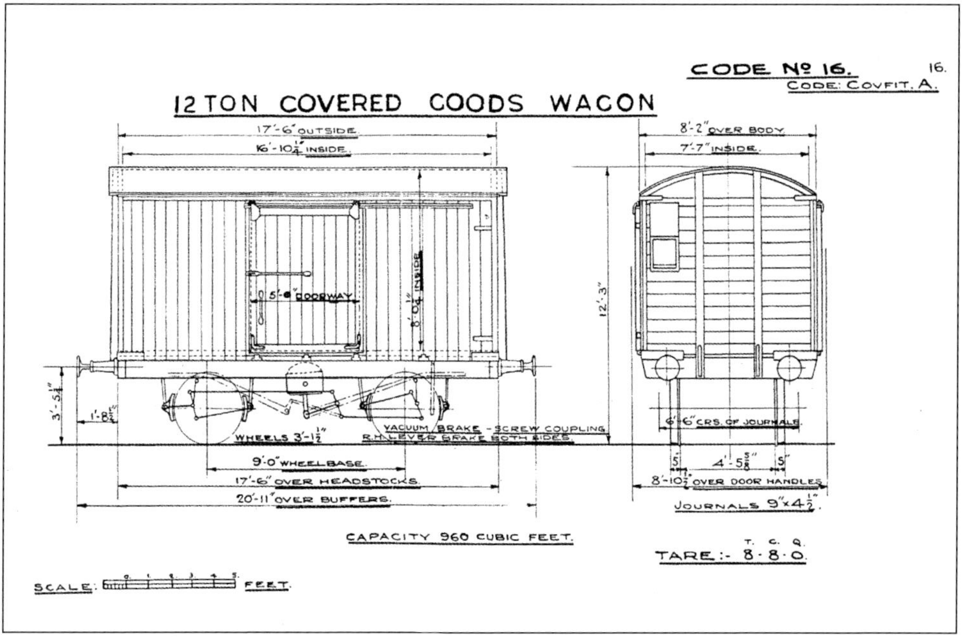

LOCOMOTIVES AND ROLLING STOCK

LNER 12 tons covered goods wagon to code 16.

LNER 12 tons covered goods wagon with steel ends to code 171.

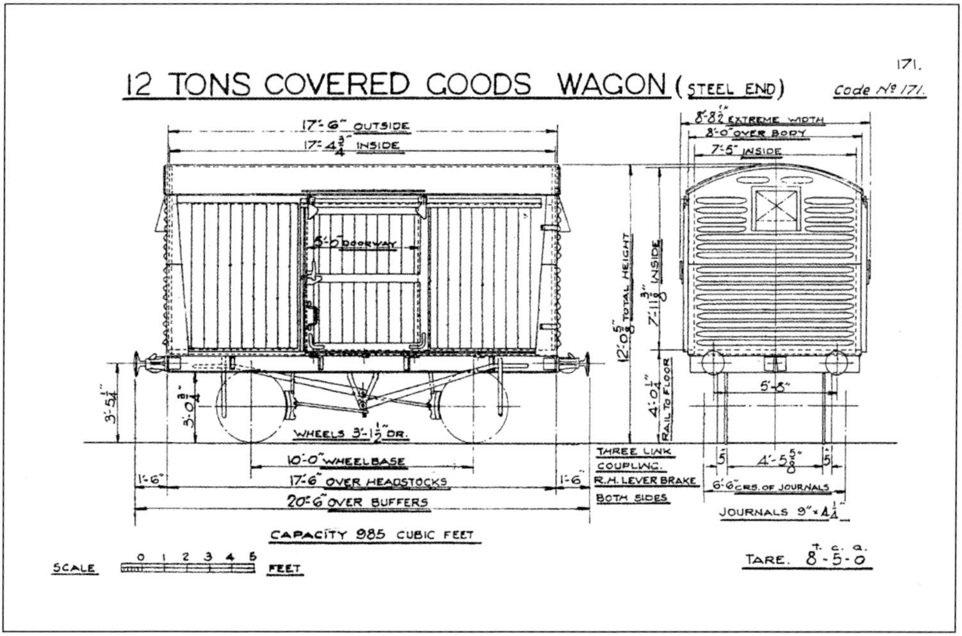

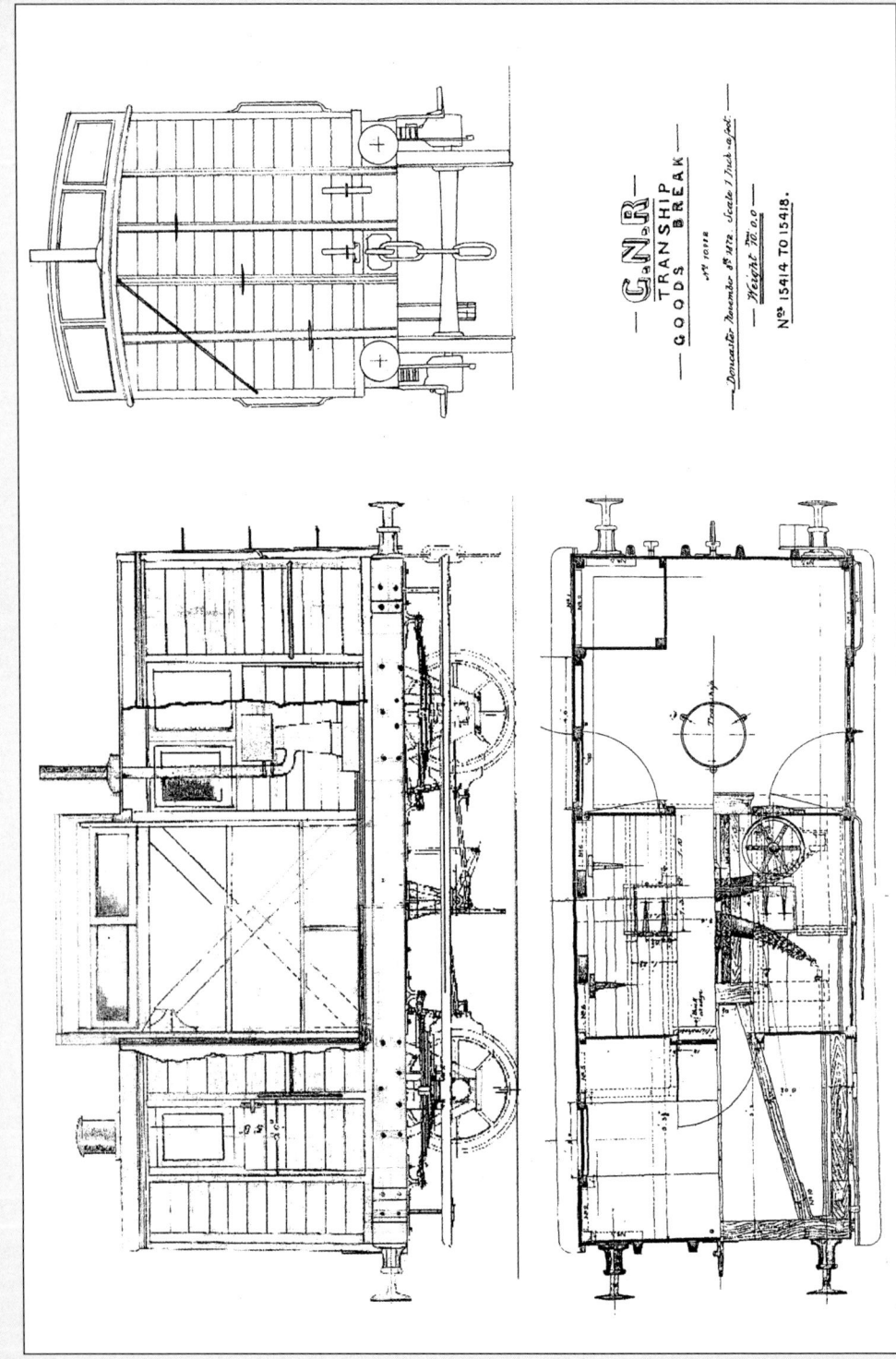

in. over headstocks and with a 9 ft 0 in. wheelbase. All were used on vegetable and sugar beet traffic and for general merchandise. Fitted and unfitted 12 ton capacity, 9 ft 0 in. wheelbase covered vans to code 16 conveyed perishable goods, fruit and malt, and later some were designated for fruit traffic only. From 1934 12 ton capacity vans to code 171, with steel underframes and pressed steel corrugated steel ends were introduced whilst at the same time the wheelbase was extended to a length of 10 ft 0 in. Specific fruit vans with both 9 ft 0 in. and 10 ft 0 in. wheelbase also saw service on the Ramsey North branch for malt traffic. Agricultural machinery was conveyed on 'lowfit' wagons. LNER brake vans provided for the branch included, 20 ton 'Toad B' to code 34 and 'Toad E' to code 64 vehicles, with 10 ft 6 in. wheelbase and measuring 22 ft 6 in. over buffers. Later 'Toad D' brake vans to code 61 with 16 ft 0 in. wheelbase, and measuring 27 ft 5 in. over buffers were employed. After nationalization many of the older wooden vehicles were scrapped and much of the traffic conveyed in open wagons was carried in the standard 16 ton all-steel mineral vehicles.

In GNR days the body, solebar and headstocks of ordinary goods wagons were painted oxide brown, sometimes called 'milk chocolate', whilst the ironwork below solebar level, buffer guides, buffers, drawbar plates and couplings were black. Lettering was white. The LNER wagon livery was grey for non-fitted wagons and covered vans, whilst all vehicles fitted with automatic brakes, including brake vans were painted brown red oxide, which changed to bauxite around 1940. Similar liveries were carried in BR days.

In the event of the derailment of wagons on the branch, re-railing ramps were available for use at Holme and Ramsey. These ramps used by station and shunting staff were provided to obviate the unnecessary call-out of the breakdown train. Ropes were also used to help re-rail itinerant wagons.

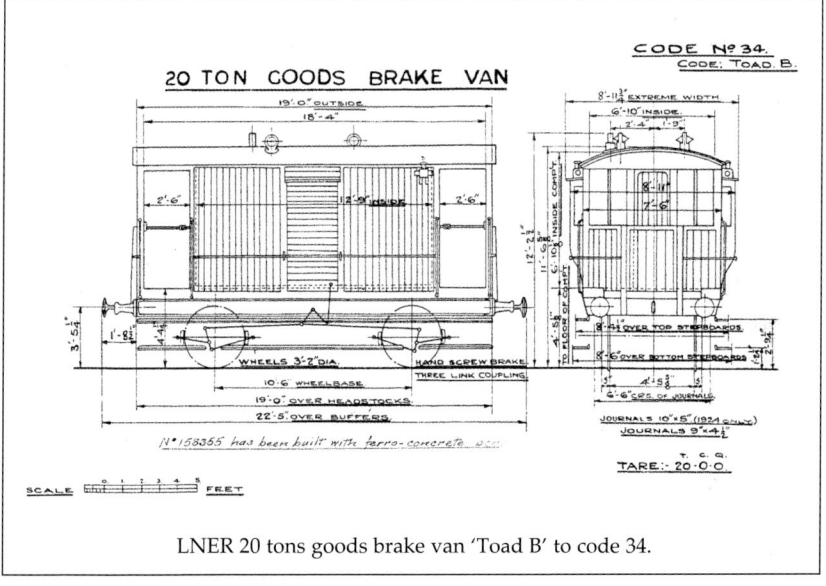

LNER 20 tons goods brake van 'Toad B' to code 34.

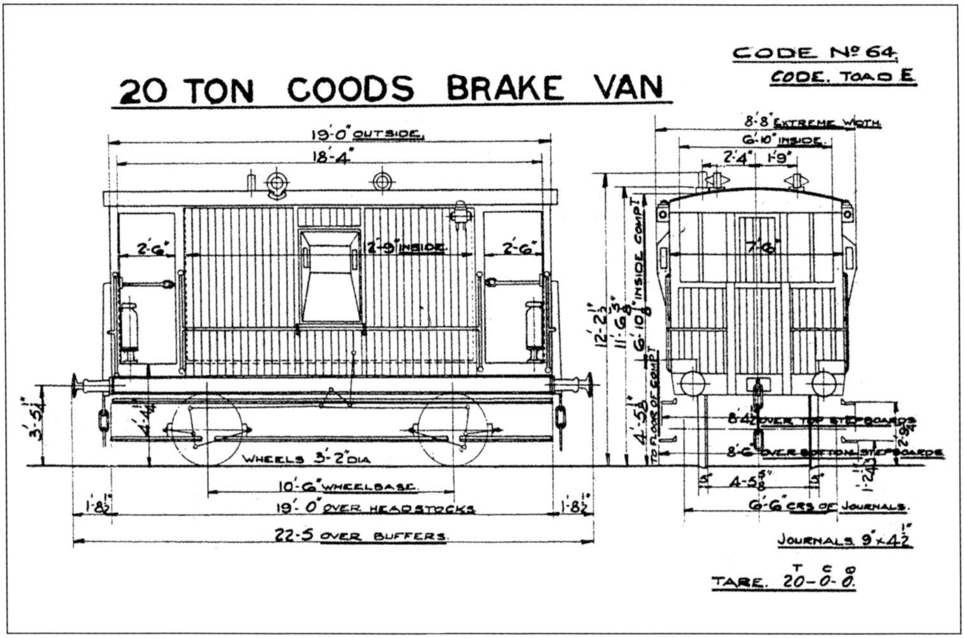

LNER 20 tons goods brake van 'Toad E' to code 64.

LNER 20 tons goods brake van 'Toad D' to code 61.

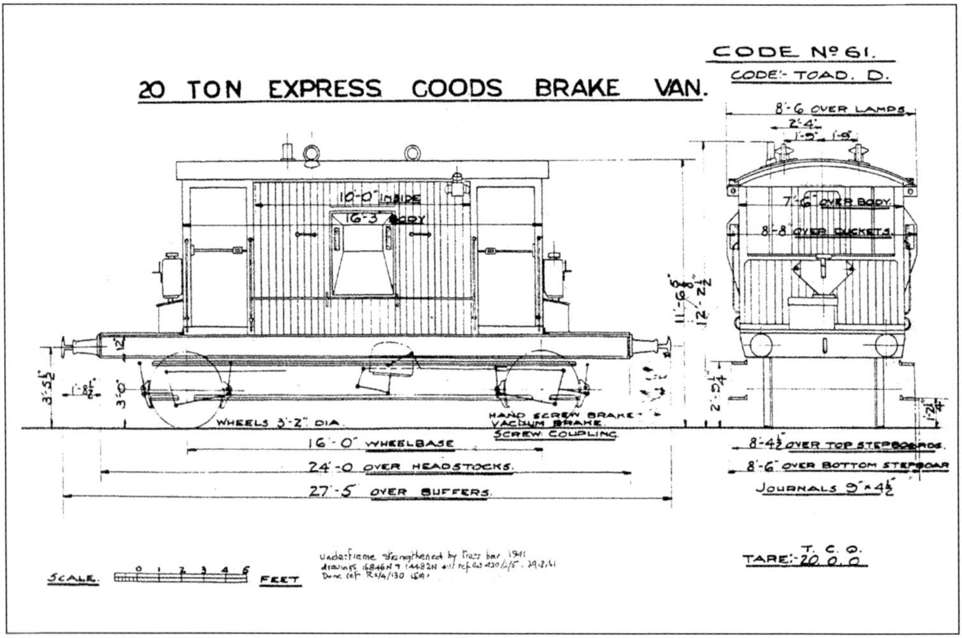

Appendix One

Lengths of Platforms, Sidings etc.

Station	Mileage ex-Holme m. c.	Platforms length in in feet	Run round loop, length in feet	Sidings length in feet	
Holme	0 00	Down Main 360	Loop 1,000	Down No. 1	300
		Up Main 305		Down No. 2	250
		Up Back 280		Down No. 3	200
				Wharf	470
				Coal Road	250
				Shed Road	300
				Shunt Neck Reception	700
				Up No. 1	2,554
				Up No. 2	871
				Up No. 3	844
				Up No. 4	316
				Up No. 5	238
				Up No. 6	283
St Mary's	3 62	Down side 150		Back Road	541
				Dock Road	304
				Loop Road	788
Ramsey	5 65¼ mileage to end of branch	Up side 240	Loop 575	Coal Road	470
				Shed Road	780
				Middle Road	760
				Engine Shed	160
				Cart Road	440
				Granary	680

Appendix Two

Bridges and Culverts

No.	Location	Mileage from King's Cross (m. ch.)	Name	Under or over	Type	Spans	Square span between abutments or supports (ft in.)	Skew span between abutments or supports (ft in.)	Construction
–	Holme & St Mary's	69 48	Culvert	Under			2 0		Wood.
–	Holme & St Mary's	69 48	Culvert (2)	Under			1 6		Wood under gateway.
–	Holme & St Mary's	69 66	Culvert (2)	Under			1 6		Wood under gateway.
–	Holme & St Mary's	69 73	Culvert (2)	Under			2 0		Wood under gateway.
–	Holme & St Mary's	70 02	Culvert (2)	Under			2 4		Wood under gateway.
–	Holme & St Mary's	70 14	Culvert (2)	Under			3 0		Wood under gateway.
–	Holme & St Mary's	70 20	Culvert (2)	Under			3 0		Wood under gateway.
–	Holme & St Mary's	70 32	Culvert	Under			6 3½		Wood, superstructure removed July 1975.
–	Holme & St Mary's	70 58	Culvert (2)	Under			1 6		Wood under gateway.
–	Holme & St Mary's	70 69	Culvert	Under			4 6		Wood, superstructure removed July 1975.
–	Holme & St Mary's	71 34	Culvert (2)	Under			2 0		Wood under gateway.
–	Holme & St Mary's	71 35	Culvert	Under			4 0		Wood, superstructure removed July 1975.
–	Holme & St Mary's	71 60	Culvert (2)	Under			1 6		Wood under gateway.
–	Holme & St Mary's	72 04	Culvert	Under			5 8½		Wood, superstructure removed July 1975.
–	Holme & St Mary's	72 07	Culvert	Under			2 0		Wood under gateway.
1	Holme & St Mary's	72 23½	Nightingale Corner	Under	Public towing paths	3	29 6	31 10 Eastern 31 10 Centre 32 7 Western	Cast iron girders on timber piles 11 ft 4 in. creosoted deal braces and binder fixed 1905 TR. Demolished 1975.
–	Holme & St Mary's	72 43	Culvert (2)	Under			2 0		Wood under gateway.
2	Holme & St Mary's	72 62½	Hogmere Drain	Under	Drain	3	7 11 11 7 8 1	Eastern Centre Western	Timber LR. Demolished July 1975.
–	Holme & St Mary's	72 78	Culvert (2)	Under			2 0		Wood under gateway.
–	St Mary's & Ramsey	73 26	Culvert	Under			2 0		Wood under gateway.
–	St Mary's & Ramsey	73 52	Culvert	Under			2 0		Wood under gateway.
–	St Mary's & Ramsey	73 57	Culvert	Under			4 2		Wood, demolished July 1975.
–	St Mary's & Ramsey	73 72	Culvert (2)	Under			2 0		Wood under gateway.
–	St Mary's & Ramsey	74 01	Culvert (2)	Under			2 0		Wood under gateway.
–	St Mary's & Ramsey	74 56	Culvert	Under			4 0		Wood, demolished July 1975.
–	St Mary's & Ramsey	74 68	Culvert (2)	Under			2 0		Wood under gateway.
–	St Mary's & Ramsey	74 78	Culvert	Under			2 0		Wood under gateway.

Appendix Three

Level Crossings

No.	Location	Mileage from King's Cross m. ch.	Local name	Status	
1	Holme & St Mary's	69 45		Occupation	12 ft field gates
2	Holme & St Mary's	69 45		Occupation	12 ft field gates
3	Holme & St Mary's	69 66		Occupation	5 bar gate
4	Holme & St Mary's	69 66		Occupation	5 bar gate
5	Holme & St Mary's	69 72		Occupation	12 ft field gates
6	Holme & St Mary's	69 72		Occupation	12 ft field gates
7	Holme & St Mary's	70 02		Occupation	12 ft field gates
8	Holme & St Mary's	70 02		Occupation	12 ft field gates
9	Holme & St Mary's	70 14		Occupation	12 ft field gates
10	Holme & St Mary's	70 14		Occupation	12 ft field gates
11	Holme & St Mary's	70 20		Occupation	12 ft field gates
12	Holme & St Mary's	70 20		Occupation	12 ft field gates
13	Holme & St Mary's	70 34	Whittlesea Mere	Public	Gates opened by gatekeeper
14	Holme & St Mary's	70 57		Occupation	12 ft field gates; closed in 1947
15	Holme & St Mary's	70 74	Triangle	Occupation	12 ft field gates
16	Holme & St Mary's	71 35	Robinson's	Occupation	12 ft field gates
17	Holme & St Mary's	71 51	Long Drove	Public	Gates opened by gatekeeper
18	Holme & St Mary's	71 60	Townsend	Occupation	12 ft field gates
19	Holme & St Mary's	71 74	Wade's	Occupation	12 ft field gates
20	Holme & St Mary's	72 07		Occupation	12 ft field gates
21	Holme & St Mary's	72 43	Bunnage's	Occupation	12 ft field gates
22	Holme & St Mary's	72 77	St Mary's West	Occupation	12 ft field gates
23	St Mary's & Ramsey	73 14	St Mary's station	Public	Gates opened by gatekeeper
24	St Mary's & Ramsey	73 52		Occupation	12 ft field gates; closed January 1956
25	St Mary's & Ramsey	73 72	Shelton's	Occupation	12 ft field gates
26	St Mary's & Ramsey	74 02	Smallholders'	Occupation	12 ft field gates
27	St Mary's & Ramsey	74 27	Gull's	Occupation	12 ft field gates
28	St Mary's & Ramsey	74 27	Gull's	Occupation	12 ft field gates
29	St Mary's & Ramsey	74 69	School Farm	Occupation	12 ft field gates
30	St Mary's & Ramsey	74 78	School Fen	Occupation	12 ft field gates

Acknowledgements

The publication of this history would not have been possible without the help of many people who have been kind enough to assist. In particular I should like to thank:

The late A.R. Cox
The late W. Fenton
The late G. Parslew
P. Townend
The late G. Woodcock
The late Dr I.C. Allen
The late Canon C. Bayes
The late Peter Proud
The late Eric Neve
The late M.T. Heugh
The late E. (Ted) Nye
Tim Hatton
M. Brooks
Peter Webber
Colin Holmes
J.E. James
Eric Marshall
The late K. Lake
The late Michael Back
Robert Powell

and the many other active and retired railway staff, some of whom worked on the Ramsey North branch, also former footplate staff at New England depot.

Thanks are also due to:

National Archives
British Rail, Eastern Region
The House of Lords Record Office
The British Library Newspaper Library
County Record Office, Cambridge
County Record Office, Huntingdon,
Huntingdon Library
Cambridgeshire Library
Peterborough Railway Society/Nene Valley Railway
Great Eastern Railway Society
Great Northern Railway Society

Bibliography

General works

Allen C.J, *The Great Eastern Railway*, Ian Allan
Bird G.F, *Locomotives of the Great Northern Railway*, Loco Publishing Co.
Gordon D.I., *Regional History of the Railways of Great Britain - Vol 5 - Eastern Counties*, David & Charles
Gordon W.J., *Our Home Railways*
Grinling C.H., *History of the Great Northern Railway*, Methuen
Groves N., *Great Northern Locomotive History*, RCTS
Joby R.S., *Forgotten Railways of East Anglia*, David & Charles
Nock O.S., *Great Northern Railway*, Ian Allan
RCTS, *Locomotives of the LNER*
RCTS, *Locomotives of the GNR*
Tatford B., *Story of British Railways*
Wrottesley J., *The Great Northern Railway*, Batsford

Periodicals

Bradshaw's Railway Guide
Bradshaw's Railway Manual
British Railways (Eastern Region) Magazine
Buses
Great Eastern Railway Magazine
Herapath's Journal
Huntingdonshire Life
Locomotive Carriage and Wagon Review
LNER Magazine
Railway Magazine
Railway Times
Railway World
Trains Illustrated

Newspapers

Cambridgeshire Chronicle
Cambridge Independent Journal
Huntingdonshire County Guardian
Huntingdon, Isle of Ely, Bedford, Peterborough and Lynn Gazette
Huntingdonshire Post
Wisbech Advertiser

Also Minute Books of the Eastern Counties Railway, Great Eastern Railway, Great Northern Railway. Great Northern & Great Eastern Joint Committee, London & North Eastern Railway

Working and public timetables - GNR, GER, LNER and BR(ER)
Appendices to working timetables - GNR, LNER and BR(ER)
Miscellaneous working instructions - GNR, GER, LNER and BR(ER)

Index

Accidents and incidents, 15, 16, 23, 34, 36-7, 39-42, 48-9, 64-5, 93, 94, 155-6
Acts of Parliament, 9, 22, 24, 28, 30, 31, 33, 34, 36, 38, 42, 44, 48, 55, 93
Benwick branch, 48, 73
Birt, William, 37, 38, 42, 43, 93
Board of Trade (including inspections), 17, 18, 37 et seq., 89, 93, 121
British Railways, 63, 67, 73, 121, 149, 154
Bus services, competition, 53, 55, 57, 60
Bus services, replacement, 59, 61, 63, 74, 109, 111
Chatteris, 7, 14, 17, 116
Closure to goods, 5, 73, 95, 114, 149
Closure to passengers, 62-3, 163
Cutting of first sod, 13
Darlow, Thomas, 9, 14
Diesels introduced, 67, 149
Eastern Counties Railway, 5, 7, 8, 14
Ely, 6
Ely & Huntingdon Railway, 7
Engine loads (summary), 120
Fellowes, Edward, 8, 9, 11, 13, 15, 17, 18, 19 et seq., 27 et seq., 33, 35
Fox, Sir Charles, 11, 13, 16 et seq., 20
Goods facilities (summary), 119
Great Eastern Railway, 5, 14 et seq., 20 et seq., 27 et seq., 33 et seq., 42 et seq., 48, 51, 53, 55, 93, 97, 100, 106, 146, 163
Great Northern & Great Eastern Jt Line (incl. Committee), 35, 37, 45, 99
Great Northern Railway, 5, 7 et seq., 19 et seq., 33 et seq., 41 et seq., 48 et seq., 55, 63, 75, 89 et seq., 100 et seq., 115, 118, 119, 121, 124, 125, 129, 133, 134, 139, 141, 143, 146, 147, 153, 156, 161, 163, 169
Headlamp codes, 153
Holme, 5, 7 et seq., 22 et seq., 30, 34, 38 et seq., 44, 45, 48, 49, 57 et seq., 63 et seq., 74, 75, 79, 88, 89 et seq., 100 et seq.,134, 141, 146, 147, 151 et seq., 163, 169, 171 et seq.
Holme suspension bridge, 34-5, 75
Huntingdon, 5, 6, 19, 22, 25, 35, 53, 57, 59, 60, 96, 97, 100, 105, 115 et seq., 146, 147
James, Thomas, 11
Lease by GNR, 31, 42 et seq., 51, 93
London & North Eastern Rly, 5, 55 et seq., 74, 89, 94, 96, 107, 111, 115, 116, 120, 121, 134 et seq., 143 et seq., 153, 154, 163, 165, 169
March, 7, 14, 22, 33, 35, 60, 151
Needingworth Jn, 35
New England shed/yard, 48, 67, 97, 103 et seq., 123, 133, 134, 137 et seq., 143 et seq., 153

Northern & Eastern Railway, 7
Oakley, Sir Henry, 28, 29, 31, 34, 36 et seq.
One Engine in Steam working, 17, 43, 93
Opening of railway, 18
Parkes, Charles, 35, 36
Partial passenger withdrawal, 5, 59, 109, 111, 117
Partial purchase by GER, 20-1, 29
Peterborough, 5, 6, 15, 22, 23, 25, 33, 34, 36, 38, 45, 48, 53, 60 et seq., 73, 74, 89, 91, 95, 97, 99, 100 et seq., 124, 125, 129, 134, 141, 145, 149, 151, 153, 154, 157, 165
Pre-opening special train, 18
Ramsey & Somersham Jn Rly, 5, 33 et seq., 42, 43, 45, 100
Ramsey gas works, 9, 14, 117
Ramsey High St (later East), 38, 43, 55, 59, 65, 67, 96, 117, 118, 134
Ramsey North, 13, 16 et seq., 30, 35 et seq., 45, 49, 55, 59 et seq., 67, 73, 74, 83, 88, 89 et seq., 100 et seq., 123, 137, 141 et seq., 149, 163, 169, 171 et seq.
Ramsey Railway, 5, 9, 13 et seq., 19 et seq., 27, 33, 35, 37, 74, 89, 91, 116
Ramsey (town), 6 et seq., 14, 15, 19 et seq., 33, 49
Route availability, 121, 134, 149
St Ives, 5 et seq., 11, 13 et seq., 20, 22, 33 et seq., 55, 57, 59, 60, 116, 117
St Mary's, 11, 13, 15 et seq., 19, 30, 37, 38, 40, 43, 45, 49, 60, 63, 67, 73, 74, 79, 83, 88, 89 et seq., 100 et seq., 134, 146, 163, 171 et seq.
Serjeant, Frederick, 29, 30, 33
Simpson & Walker, 8, 13
Simpson, W.S (contractor), 13 et seq.
Somersham, 7, 8, 16, 20 et seq., 27, 33, 60, 65, 73, 96, 97, 100
Somersham Ramsey & Holme Rly, 8, 13, 14
Speed limit, 88, 113
Stafforth W., 21, 22, 24, 25, 29, 30
Strikes, 53, 55
Sugar beet traffic, 60, 73, 118
Sunday train service, 111-2
Swarbrick, Samuel, 29, 31, 34
Takeover by GER, 31
Train Staff and Ticket working, 93
Warboys, 7, 8, 11, 14, 33, 59, 65, 73, 96
Whistle codes, 154
Whittlesea, 33, 55, 60, 116, 146
Wisbech, St Ives & Cambridge Jn Rly, 7
Working agreement with GNR, 13, 24, 25
World War I, 51, 61, 117, 118, 137, 143, 145
World War II, 61-2, 75, 89, 111, 116 et seq., 137, 163